Dear

I trust you will find concepts in this book that you will be able to apply to your own safety program to further promote the health and safety of your workers at U.S. Steel Gary Works.

Richard Hults?

CONSTRUCTION SITE SAFETY

A Guide for Managing Contractors

Richard D. Hislop

LEWIS PUBLISHERS

Boca Raton London New York Washington, D.C.

Library of Congress Cataloging-in-Publication Data

Hislop, Richard D.
    Construction site safety : a guide for managing contractors /
Richard D. Hislop.
      p.   cm.
    Includes bibliographical references and index.
    ISBN 1-56670-304-2
     1. Building--Safety measures. I. Title.
    TH443.H57 1999
    690¢.22—dc21   99-22938
                                                                       CIP

© 1999 by CRC Press LLC
Lewis Publishers is an imprint of CRC Press LLC

No claim to original U.S. Government works
International Standard Book Number 1-56670-304-2
Library of Congress Card Number 99-22938
Printed in the United States of America 1 2 3 4 5 6 7 8 9 0
Printed on acid-free paper

Preface

This book presents basic concepts and tools for the practical development of contractor safety programs from project conception to completion. It is intended to be a ready reference for facility owners who wish to institute an effective contractor control program. It should also be of value to the conscientious contractor who wishes to be responsive to customers who expect their contractors to implement safe work practices. Architects and engineers who will design structures, construction managers who oversee projects, superintendents who supervise various phases of the work, and the safety professional who, as a member of the planning team, will be responsible for providing the guidance necessary to assure that safety is incorporated into all phases of the construction project will also find this to be a useful guide.

Concern regarding management of the safety and health aspects of contracted work has increased in recent years. While this concern has been predominately focused on high-risk environments such as construction and petrochemical industries, similar issues are present in other contracted work environments. The amount of work being performed by service contractors is increasing, in response to greater outsourcing of cyclical, undesirable and high-risk work by organizations that have downsized or simply wish to transfer risk. For a variety of reasons addressed in this book, contractor safety will become an even more prominent issue as the demographics of the contracted workforce employed to meet this demand continue to shift.

The principles presented in this book have been applied in a variety of geographical locations around the world. The principles have been effective irrespective of the workforce to which they have been applied, whether they were third-world national workforces overseas, direct hire, or contracted labor in the U.S.A. The same positive results as measured by low injury indices and low dollar losses have also been achieved irrespective of the project discipline, whether it was the construction of railroads, port facilities, mine maintenance facilities, airports, roads, bridge projects, multistory buildings, or residential complexes.

For years academics have expounded safety theories, but few have provided how-to guidance for the facility owner or construction person attempting to implement a safety program in a contracted work environment. During the past few years individuals such as David MacCollum and Janine Reid have begun to produce more practical safety guidance supporting the construction industry. I believe that there is a need for a practical guide for the implementation of an effective project safety program from a host employer's perspective.

At this point I don't believe there is anyone who would dispute the value of providing a safe work environment for both employees and contractor personnel. The extent to which the ideas in this book are implemented should be based on the complexity of the job and the level of risk present. Generally, the greater the risk or complexity of the work, the more detailed and comprehensive the controls should be.

This material is not meant to provide legal advice in any sense and any organization seeking to implement or improve contractor safety controls must do so in consultation and compliance with its own safety professionals and legal counsel.

I would like to express my appreciation to my colleagues in the National Safety Council's Construction Division. Their practical experience in managing contractor safety operations throughout the world has provided me with an invaluable sounding board as I developed this book. I would particularly like to thank those individuals who supported me with material, text, and moral support throughout the development of this book. Thank you, David MacCollum for the hours you spent discussing safety philosophies and safety planning concepts; Phil Colleran, your in-depth knowledge of safety and regulatory standards has been a reliable and ready resource of information for me; Frank Keres, your contract knowledge and humor helped liven up an otherwise dull but very important subject; and Janine Reid, your willingness to share your detailed knowledge of emergency and crisis management and help me develop an otherwise complex subject into a practical guide for the purposes of this book. Each of these individuals whether he or she realized it or not was most helpful in providing me with guidance and insight during the development of this book as well as in the management of my own business affairs.

About the Author

Richard Hislop, PE, CSP

Richard is an engineering and safety professional with broad project and safety management experience. He has been heavily involved in construction projects throughout his career, starting while serving with the Canadian Corps of Engineers, working with Exxon Corporation and most recently at Argonne National Laboratory. Although initially he began working as a field engineer, he is now experienced in estimating, cost and schedule, facility operations, facility maintenance, safety and quality assurance. Much of Richard's efforts during the past few years has been spent in the development and implementation of safety programs for organizations ranging from small construction contractors to international construction projects employing in excess of 10,000 laborers.

Richard also manages a successful safety consulting practice (Virtual Safety Inc.) with clients ranging from residential and heavy construction contractors to municipalities. He conducts training on behalf of the National Safety Council and independently for organizations seeking to improve their construction safety programs and the management of contractor work safety. Richard is the author of numerous articles on subjects related to construction safety which have been published in both professional and trade journals. He is also an active participant in the National Safety Council's Construction Division, the American National Standards Institute's A10 committee for Construction and Demolition Operations, and the American Society of Safety Engineers.

Contents

Introduction

Managing project safety can be learned. While some people seem to know intuitively how to motivate people to work safely, most effective project managers learn this skill. This book explains how to manage project safety in order to improve the odds of having an injury free work place.

Skillful safety management involves knowing what is to be done, who is to do it, and when and how it should be done. This book considers all these factors and also explains why. If project managers are to apply their judgement wisely, they need to know the rationale for each requirement as well as how to implement the requirement itself.

The program elements described in this book apply to both large and small projects, from conceptual design through construction completion. Large or complicated projects offer more opportunity to overlook critical aspects than do small or simple projects, but small and simple projects still require the application of the same fundamental safety tenets. In either case the project managers must pay attention to the factors discussed in the following chapters whether explicitly or implicitly, if they are to have a project free of injuries.

As a text on contractor safety management, this book makes no attempt to address the nuances peculiar to selected types of projects. Rather, I assume that the readers will supply their own insights into their specific situation and apply the basic principles here in a graded manner to suit their own culture.

As noted, this book is a mixture of the what, who, when, how, and why of project safety management. Some background, however, is given first so the context of these topics is clear. For this reason, Chapter 1 provides a background on the why of construction safety and is followed by an outline of basic safety program elements. Chapters 3, 4, and 5 address the roles and responsibilities of the project manager and other characters in the construction process. Chapters 6 through 8 set the stage for communicating with the contractor. In Chapters 9 through 14, we address individual safety tools and techniques that must be integrated in the execution of construction work. Chapter 15 provides guidance on what should be considered when closing a job. Finally, in the epilogue I try to show the interdependence of the tools and techniques used to implement an effective safety program.

Not everyone remembers or wishes to consider safety in their planning and project implementation. In many cases, the consequences of inept safety management are not terribly significant. The work force on a small project may experience a couple of scraped knuckles or even a lost day case. It may be the assumption of the project manager that things happen and "after all people do get hurt in construction." But if the circumstances leading up to an incident are complicated with a few more factors, the consequences may be significant; instead of a scraped knuckle, the worker could lose a couple fingers. Instead of a simple slip and fall near a leading edge, the worker might go over the edge if adequate fall protection is not in place. While the manager of a small installation project can afford to be inept, the project manager of a multistory building with numerous contractors on site does not have this luxury. Nor does any engineer who is in the position to coordinate contracted work and to a great degree influence the safety of that workforce.

Chapter 1

Safety in Construction
Value of Safety

"The first duty of business is to survive and the guiding principle of business economics is not the maximization of profit, but the avoidance of loss."

Peter Drucker,
Management Consultant

The objective of business is to make money and to be profitable; revenues must exceed expenses. That's the secret to success in all businesses, including construction. Both large self-insured corporations and "Mom and Pop" entities are bound by this formula. Both spend considerable effort attempting to increase their profitability, for example, by widening markets and increasing sales or by reducing component costs and making more efficient use of labor. However, one element that is often overlooked that has a negative effect on net profits is the cost of accidents. Direct costs, resulting from accidental equipment damage or personal injuries, plus the associated indirect financial impacts resulting from schedule disruptions, increases in insurance and workers compensation premiums, can be devastating to profits. As more organizations involved in construction activities look for ways to reduce project-associated costs, they are beginning to recognize that unsafe work practices cost them money.

Catastrophic construction-related incidents that are the subject of front-page news, such as the high-rise hoist collapse at Times Square on July 21, 1998, when 300 feet of hoist mast was hurled into the street and building below, killing an 85-year-old woman, or the sinkhole that appeared in the streets of Los Angeles the year before caused by tunneling operations, are not the sort of occurrences that are the source of the majority of losses suffered by constructors.

The losses that are crippling the construction industry and those that most seriously impact the majority of its workers are the multitude of minor injuries that occur on a regular basis. The hand laceration that may only cost an insurance carrier $600 lessens the worker's efficiency and will eventually cost the employer $1,500 to $2,500 in increased insurance premiums. A lower-back injury with a claim value of $2,000 will cost an employer approximately $34,000 in increased insurance premiums, legal fees, and settlement costs. Additionally, this sort of injury will potentially interrupt the employee's income stream long enough to be seriously disruptive to his family life.

Other Sources of Losses

Other sources of losses eroding the profitability of constructors are those resulting from the absence of supervisory consideration of safety in work planning and execution. This is exemplified by the situation where a well-meaning superintendent on a post office construction job with an anticipated profit of $750,000, cost his employer $1,500,000 as the result of injury claims and increased insurance premiums resulting from trying to get the job done quickly with little regard for the safety of the work crew.

The root cause of the majority of "safety-related" losses is the absence of a systematic process to identify and mitigate workplace hazards and unsafe work practices. It's the result of the failure of supervision and management to effectively communicate the importance that safety has on the continued economic viability of the organization employing them and the importance of maintaining the workforce's health so that there are qualified individuals available to do the work.

Successful projects in the future will be those that recognize that changes are taking place in construction and its workforce. Those host employers[1] who involve contractors and their own workforce in the safety process will be more successful. Management commitment at all levels of the contracted work hierarchy and manager personal involvement set the stage for successful safety programs. Management must be clearly committed to safety and must establish and communicate a vision to the project team.

A successful safety program is not established through random application of "off-the-shelf" safety initiatives. Safety must be integrated into the conduct of routine work.

---

[1]   Host Employer - This is either the Owner or the Project Constructor who controls the workplace and is in charge of the overall work performed there.

I. Introduction

Safety is far more than craftsmen wearing hard hats on construction sites. It is a philosophy that identifies and eliminates job site hazards throughout the lifecycle of a work project. It is a philosophy that discourages work practices that place individuals at risk of injury. It is the integration of safety into the daily work process. It is the promotion of an environment where each person in the project's construction hierarchy has a role and responsibility for safety.

Accident statistics show that construction is one of the most dangerous industries in the world. The National Institute for Occupational Safety and Health (NIOSH) reports[2] indicate that, on the average, between 1980 and 1993 1,079 construction workers were killed on the job each year. This is more fatalities than in any other industry. Construction-related deaths account for approximately 20% of work-related fatalities in a market segment that employs only 5% – 6% of the national labor force. Some contributing factors to this condition are obvious — incomplete structural connections, temporary facilities, tight work areas, varying work surface conditions, ever-changing work sites, multiple operations, and crews working in close proximity. However, several easily overlooked factors, such as lack of preplanning, inadequate selection of contractors, and *laissez faire* attitudes are significant contributors to these statistics.

Injury-free working environments are important to business because they eliminate financial losses associated with injury claims, lost work time, and schedule delays. As we are generally in business to make a profit, we must recognize that profits are directly related to the degree to which we are able to avoid losses.

An issue of great personal impact on owners, their representatives, project managers, and superintendents is that they can be held responsible in courts of law for injuries resulting from unsafe work practices or unsafe job site conditions, if these could have reasonably been foreseen and controlled. This is so even if the occurrence is caused by and affects only subcontractor employees, where it can be demonstrated that there was the absence of an effective safety program.

When we think of construction safety, our first thought is generally that this is the responsibility of the contractor. This focus reflects the contractor's control of their workplace and their work practices. There are, however, many more characters who must be involved in establishing a safe job site than just the contractor. All the participants in the construction process, from the client commissioning the work to the constructor and to the men and women who perform the work, are integral to the process of establishing a safe workplace. Therefore, throughout the construction process we must have a means in place by which hazards can be identified and effectively controlled, and safe work practices promoted.

---

[2] National Institute for Occupational Safety and Health, *NIOSH Research Projects,* August 1997.

II. Changes in the Industry

Job site safety and health in construction have improved in recent years, as reflected in the reduction in injury and illness rates since the early 1990s.[3] Accidents and even fatalities were at one time an accepted fact of life in the construction industry. In the 1930s, when the Golden Gate Bridge was built, it was expected that there would be one fatality for each million dollars of construction work. The estimated cost of completion of the bridge was $32 million. The project safety program was considered a success when the bridge was completed with only 17 fatalities. With time our expectations have changed. It is not uncommon now to hear of billion dollar construction projects where millions of man-hours were worked without a single lost time case.

CONSTRUCTION FATALITY RATE

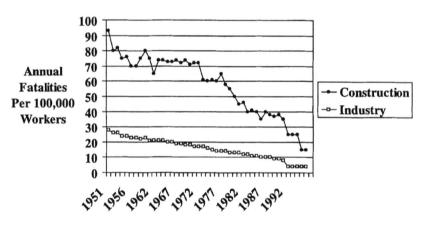

Perhaps the most significant factor attributed to this reduction in deaths is increased management commitment to safety. Management has been motivated to make a greater commitment to safety based on their growing realization of the impact that increasing costs of worker's compensation payments, higher dollar value judgments in lawsuits, and increased OSHA penalty structures have on profitability. And their corporate image.

Though there is evidence of a general trend toward improved job site safety, a great disparity still exists within the construction industry. While many facilities and projects have worked millions of construction man-hours without a

---

[3] U.S. Department of Labor, Bureau of Labor Statistics, *Safer Construction Workplaces Evident During the Early 1990s*, January 1995.

single lost time injury, many contractors have injury and insurance rates two or three times higher than the industry average.

III.   Industry Trends and Challenges

Safety in the contracted work environment is under pressure by a variety of industry trends. Downsizing and outsourcing of work, the increasing complexity of operating systems, increased specialization of equipment, and more potent chemical products create an environment ripe for accidents.

It is an economic reality that most organizations cannot afford the luxury of retaining a full complement of employees with the technical expertise to attend to the construction, modification, maintenance, as well as the operation requirements, of the complex systems common to today's operations. As a result, specialty contractors are generally employed to perform construction work and to meet specialized cyclical work requirements.

When facility and construction company owners are asked to identify the greatest daily challenges with which they must deal, they are generally contractor-related issues. Among the greatest problems faced by managers are finding qualified contractors, dwindling productivity, getting quality work done, and the reluctance of some contractors to establish and implement safety programs. The reasons for these problems lie in several areas.

Availability of experienced, qualified workers is at a 30-year low. This is perhaps one of the greatest problems contractors face as they attempt to staff their jobs. The construction industry is faced with an aging workforce and will have to recruit and train as many as 250,000 craft workers per year to replace those that leave. Young people appear to be disinclined to want to work in the trades. Unions and organizations such as the National Electrical Contractors Association and the National Center for Construction Education and Research are addressing this problem in an attempt to increase the workforce. In fact, this is such a concern that the Business Roundtable recommended in 1997 that owners should only do business with contractors who invest in training and maintaining the skill of the workforce. The decline in the availability of trained craftsmen means that contractors have the choice of either paying bonuses and higher wage rates to attract qualified workers or hiring less experienced workers to get their work done. The presence of new and unskilled individuals in the construction work environment presents an increase in the potential for accidents and the likelihood of injuries.

Cyclical nature of construction work and relatively short job duration, such as "10-day smelter outages," are difficult to staff. The cyclical demand for contracted services generates financial and safety problems for both the contractor and the host employer. "Boomers", as some of the travelers are called, will only work on utility and plant maintenance projects if they are paid for six or seven 10s plus per diem expenses, because they are supporting two residences. Contractors have little choice but to accommodate them in the interest of

5

securing the labor skills needed to get the work done. This pressured work environment created by long work hours, resulting in inattention due to exhaustion and the accommodations that both host employers and contractors feel they need to make to secure the workforce needed to perform the work, results in high risk factors.

Industry rightsizing (downsizing) is resulting in the increased use of contracted services to perform the less desirable and often more hazardous tasks. Contractors are under significant pressure to work quickly. This combination, unless moderated with an effective safety program, places contractor employees at great risk of personal injuries and generally they do experience more injuries than do direct hire employees.

Harnessing technology in the construction industry has lagged behind that of other industries. The construction industry has also been slow to improve the efficiency of its work practices or to speed up construction processes. This failure to embrace technological advancements that could improve job site safety, such as load moment indicators and proximity alarms in cranes, or the unwillingness of iron-worker connectors to tie off in spite of the availability of new retractable lanyards, is protracting the establishment of safe job sites. However, progress is slowly being made in this area as project management considers alternatives such as assembling roof structures on the ground and lifting them into place to minimize hazards of work at heights.

Litigation associated with the construction industry is increasing. The escalating trend of workers turning to litigation to resolve injury claims and the increased cost of associated settlements and awards is inducing risk managers to attempt to transfer risk rather than promote a process of eliminating hazards and the source of the problem. No contract legalese seems to have been effective in preventing lawsuits from being filed. Lawsuits do not always result in large payments, but dealing with lawsuits is costly.

IV.   Involvement in Safety

Facility owners have a vested interest in assuring that an effective construction safety program is in place on their job sites, as the owners ultimately pay for all losses. Owners must assure that the contractors they retain to perform work implement sound safety programs that integrate safety into routine work practices.

Reasons given in the past by field supervision for poor safety performance on construction sites are no longer acceptable positions:

- "Construction is no place for sissies."
- "I don't have money for frills like safety."
- "I am forced to choose between production and safety."

Employers did not begin the process of establishing safety programs with the advent of OSHA. Responsibility to control hazardous conditions has been well documented for some time. In fact, King Hammurabi of Babylonia in approximately 2200 B.C. is credited with the following legislation:

> If a builder builds a house for a man and does not make his construction firm and the house he has built collapses and causes the death of the householder, the builder shall be put to death.
>
> If it causes the death of the son of the owner of the house – the son of that builder shall be put to death.
>
> If it causes the death of a slave of the owner of the house – the builder shall give to the owner of the house a slave of equal value.
>
> If it destroys property, the builder shall restore whatever was destroyed, and shall rebuild the house at his own expense.

Timidity born out of fear of incurring liability for contractor substandard practices by becoming involved in safety is not a sound reason to avoid taking action. Owners may well incur greater liability if they do not become involved. Therefore, development of a sound program and sensible documentation is by far a better posture than being the defendant in a lawsuit.

Who better than the party with "purse-strings authority" to affect compliance with contract provisions for safety and health? The root cause of poor safety performance is often the lack of a clear message from the owner and contractor management that safety is important. Although managers may profess the importance of safety, their actions may lead their employees to doubt them. *"Right, boss, I hear what you are saying, but I know what you really want. You want me to get the job done as soon as possible."* Where senior management may in fact be strongly committed to safety, supervisory level personnel may well be the point of disconnect between management's commitment to safety and the regular application of safe work practices by workers.

Prior to starting a new work project, the host employer must establish that safety will be an integral part of the project and must require that each employer that comes to the site have a safety program in place. The host employer must determine once work has begun that a safety process is being implemented.

Program standards, implementation criteria, and monitoring responsibility must be clearly defined at the onset of contracted activities. The contract must define that contractor and subcontractors have a contractual responsibility to perform all on-site operations in accordance with an approved site-specific environmental safety and health plan and have individuals competent to implement the program.

7

V.   Risks of Using Contracted Services

It is generally recognized that specialists can perform work in less time and with greater efficiency than can non-specialized individuals. There is a significant investment associated with training and keeping individuals abreast of the rapid technological developments in electronic and mechanical systems, and of the evolution of chemical products in today's market. The return-on-investment in training individuals to perform specialized tasks, and to keep them current so that they are considered to be technically competent, cannot be justified, unless those individuals are kept regularly occupied. Therefore, contractors are often hired to perform cyclical work that the host employer's regular employees do not have the technical competence to perform or have little interest in performing. Additionally, some managers view the use of specialty contractors as a means of transferring risk and insulating themselves from potential injuries associated with performing unfamiliar or high-risk tasks.

It is generally assumed that specialty contractors have the requisite technical knowledge and skills to perform the work for which they are retained. Host employers expect specialty contractors to perform their work competently and in compliance with applicable environment, safety, and health regulations. Although many specialty contractors do understand the risks associated with their work and perform their work with due regard for safety, and are concerned for the health of their employees, there are many who fail to take appropriate steps in this regard. Whether out of ignorance or a misguided sense of urgency, many contractors do not adequately control the hazards generated by their own work processes or fail to consider that it is their responsibility to protect other individuals who might be in the vicinity of their work. Occasionally, as the result of this lack of awareness or knowledge of the potential hazards of the products used with their work, contractors fail to protect the health of their own employees.

The following is an example of this situation. A contractor preparing a concrete floor of a dry storage room for painting was removing grease and foreign material with a cleaning agent. The workers were then going to abrade (sand) the floor to improve paint adhesion. They were having some difficulty removing a particular spot so they decided to use acetone and an electric floor buffer to remove it. The "specialty painting contractor" failed to recognize the hazards of using acetone to clean the floor in conjunction with an electric buffer. The acetone concentration in the unventilated room quickly reached the lower explosive limit for acetone. Had the individuals not had the "good sense" to be wearing respirators, they could well have been in imminent danger of asphyxiation before the flammable liquid fire/explosion occurred, due to the concentrations of the fumes in the room. The manufacturer's label on the floor buffer read "Warning, risk of fire — do not use flammable or combustible liquid to clean floor — risk of explosion. Floor sanding can result in an explosive mixture of fine dust and air. Use floor sanding machine in a well-ventilated area." The explosion resulted in structural damage to the walls of the room; one painter received second-degree burns and the other two painters sustained third degree burns.

Contractors may also fail to ensure that their employees are appropriately trained or have requisite baseline medical tests. Another oversight of many specialty contractors is their failure to inform their host employer of hazardous material they plan to bring into the workplace.

VI.   Compliance Expectations

Host employers often assume that it is the contractor's responsibility to comply with relevant environmental, safety, and health regulations. They rationalize that it is the contractors' problem if they are caught violating regulatory requirements. Host employers expect contractors to be aware of the hazards associated with their work and to take necessary precautions to protect their own employees from hazards as well as any other individuals who might be nearby and possibly affected by the hazards.

Contractor failure to comply with safety requirements may also have a detrimental effect on a facility's own operations. Occurrences resulting from contractor oversights may affect the facility's ability to continue to operate efficiently. This may, in turn, affect the completion of the work as scheduled. This was the case in an oil refinery explosion in Texas in 1997 where a piping subcontractor failed to follow the facility's electrical lockout/tagout procedures and precipitated a major fire at that location and resulting in several fatalities.

VII.   Who Pays

Another very important consideration regarding the amount of trust bestowed on a contractor's safe work practices is the cost associated with accident losses and the financial effect an occurrence may have on ongoing facility work activities. Contractor safety lapses, however minor, will have some negative effect on the host employer. Host employers often fail to realize that it is they who pay the contractor's costs as long as the contractor remains in business.

Host employers must verify that contractors are aware of the potential hazards present in the existing work site and that the contractors are prepared to control the hazards associated with their own work. This requires that the host employer has a basic understanding of the work to be performed by the contractor and its associated hazards. For example, the host employer must be advised by the contractor of such things as chemicals to be used, the hazards associated with their use, the effects of the chemicals on personnel and the symptoms that would indicate that there might be a problem. It is, after all, the host employer's workplace. Contractor-generated hazards could potentially affect the host employer's staff and visitors. Additionally, any mishaps or work delays could affect the project's completion, which in the final analysis, impacts the host employer's operation.

VIII. Establishing a Safety Program

A more structured safety program is required when working with contractors than is required for a static workforce. A static workforce generally receives more training and has better-defined safe work practices than do contractor employees. The growing trend to employ contractors to perform high-risk tasks and the increasing complexity of projects that require greater specialized work procedures call for more attention being given to ensuring that safety is integrated into their work practices.

Recognition of this fact is reflected in relatively recent additions to OSHA legislation such as "Process Hazards Management of Highly Hazardous Chemicals" 29 CFR 1926. This standard requires that owners and job-site managers[4] inform employees and contractors of the hazards to which they might be exposed. In these environments the job-site manager is also required to ensure that contractor employees are adequately trained to work safely in their new environment and that they are aware of applicable safety rules and emergency plans.

IX. Planning

Planning is a significant factor, accounting for the difference between those owners and contractors with exceptional safety records and those with recurring fatalities and exorbitant insurance rates. A survey of the safety and health programs of successful employers, and a review of reports on major construction catastrophes, reveals that planning which includes consideration of the sequence and methods of construction is an element missing in the project evolution process where serious incidents have occurred. Planning is *always* found in the evolution of successful projects.

X. Cost of a Safety Program

According to studies conducted by the Business Roundtable,[5] accidents account for an average of 6.9% of total project costs where effective safety programs are not in place. The cost of running an effective safety and health program is reported to be about 0.9% of total project costs. An effective safety and health program, therefore, offers a fair return on that investment.

The program elements included in this cost estimate are

- Salaries for safety, medical, and clerical personnel
- Safety meetings

---

[4] Job Site Manager – A person, firm, or corporation, i.e., the construction manager, general contractor, prime contractor or other entity, as designated in project documents, responsible for supervising and controlling all construction work performed on the project.

[5] The Business Round Table, *Improving Construction Safety Performance Report A-3,* January 1982.

- Inspections of tools and equipment
- Orientation training
- Site inspections
- Personal protective equipment
- Health programs such as hearing and respirator fit
- Miscellaneous supplies and equipment

An effective safety program is businesses' best guarantee of stemming profit drain caused by costs associated with injuries, equipment damage, and schedule delays. Costs associated with accidental losses cannot be depreciated or written off as a tax loss; therefore, they directly impact an organization's bottom line.

XI.   Cost of Not Having a Safety Program

Simple math illustrates the impact of the absence of a safety program. If a company were making a 3% profit, it would need to generate more than a third of a million dollars in sales to recoup the cost of a single worker's strain or sprain. This is based on the average direct cost of about $5,000 for a strain, plus another $5,000 in indirect costs. That's $10,000. To generate that amount of profit, the company would have to sell $330,000 in services. In times of keen competition and low profit margins, loss control may contribute more to profits than an organization's best salesmen.

Table 1.1 Sales required to cover losses

Yearly Incident Costs	Profit Margin				
	1%	2%	3%	4%	5%
$ 1,000	100,000	50,000	33,000	25,000	20,000
$ 10,000	1,000,000	500,000	**330,000**	250,000	200,000
$ 25,000	2,500,000	1,250,000	833,000	625,000	500,000
$ 50,000	5,000,000	2,500,000	1,667,000	1,250,000	1,000,000
$ 100,000	10,000,000	5,000,000	3,333,000	2,500,000	2,000,000
$ 150,000	15,000,000	7,500,000	5,000,000	3,750,000	3,000,000
$ 200,000	20,000,000	10,000,000	6,666,000	5,000,000	4,000,000

Specific costs incurred as the result of construction accidents are as follows:

A. Lost Production Time

The inefficiencies resulting from work interruption are a buried cost, but nonetheless do have a detrimental impact on the total cost of completing work.

B. Insurance Costs

Insurance industry premiums are based directly on accident-related costs that insurance companies have to bear, and 15% of workers' compensation costs in the U.S. are spent on construction injuries, which represent 5% of the national workforce.

C. OSHA Penalties

Increased enforcement of Occupational Safety and Health Act regulations and associated punitive fines are another accident-related cost.

D. Litigation Costs

Becoming powerful motivators to take safety more seriously are the increasing costs associated with accidents and the inevitable liability actions with potentially significant judgments that follow.

Large contractors and increasing numbers of owners are recognizing that the investment in a sound safety program can potentially produce a greater return-on-investment than any other discipline within construction.

XII.   Legal Requirements

Each state has its own set of workers' compensation laws and statutes that require employers to compensate employees for work-related injuries. In some locations the courts may even hold employers liable when a subcontractor employee has been killed or injured due to company negligence.

Roy G. Stoops, owner of C&S Erectors, Inc. of Noblesville, Indiana, was sentenced to four months in prison after pleading guilty to willfully violating occupational safety and health standards that led to the death of an employee. The case stems from a May 1, 1996 incident in which an employee of C&S Erectors fell to his death while laying steel decking on a roof approximately 35 feet above the ground at a site in Jonestown, Pennsylvania.

After an investigation by OSHA, the case was referred to the Department of Justice for criminal prosecution. Judge J. Andrew Smyser of the U.S. District Court in Harrisburg, Pennsylvania, ordered Stoops to pay $6,000 in restitution to the family of the deceased to cover funeral expenses. Stoops was placed on a one-year supervised release program. C&S Erectors, Inc. was also sentenced to one year of probation and was held jointly liable for the restitution.

According to OSHA records, C&S Erectors engaged in several unsafe practices at the Jonestown worksite, and prior to the accident, the project's general contractor warned Stoops and the company several times about hazards, including the failure to provide fall protection.

Employers with one or more employees are obligated by law to meet the minimum standards set by the Occupational Safety and Health Act (OSHA). All employers are bound to maintain a workplace free of recognized hazards that are likely to cause death or serious harm to their employees.

XIII. Contractor Selection

An effective safety program costs significantly less than the dollar losses for accidents that otherwise are apt to occur. Contractor attitudes toward safety range from minimal compliance to total commitment, so concerned host employers should consider contractors' past safety performances during the contractor selection process. Improvement in project safety can be accomplished most readily by hiring contractors with a record of good safety performance. When selection of contractors is limited, identifying their safety weakness as a target for the host employer's focused attention and involvement can reduce the potential for safety mishaps.

Chapter 2

Safety Programs

Paradigm Shift – Construction
Safety & Health

From . . . . .*How many fatalities will we have on this project?*
To . . . .*No accident is acceptable.*

I.   Introduction

Given the technical resources and safety information available today, there is no excuse for accidents on construction sites. Unfortunately, high-profile stories such as the collapse of the elevator tower in New York's Times Square continue to provide headline news. Less newsworthy, but just as devastating, are the isolated and avoidable construction-related fatalities that continue to occur with alarming frequency, such as the case of Luis Gomez, a 32-year-old construction worker who was buried in an excavation,[1] and the many individuals who die each year from work-related falls, electrocutions, and other hazards common to construction.

---

[1]   "Safety is an Obligation," *Engineering News Record*, July 20, 1998.

Accidents and personal injuries can be prevented through the mitigation of job site hazards and unsafe work practices. There is a significant financial benefit to employers in providing a safe working environment for employees and the contractors they hire. In addition, since it is morally unacceptable to knowingly expose individuals to conditions that could be reasonably expected to result in injuries, the law requires that employers provide a safe working environment. Included in establishing such an environment is the presence of clearly defined safety guidelines within which work shall be performed . . . A Safety Program.

II.   Due Diligence

Employers[2] must exercise due diligence in the development and implementation of their safety programs. That is to say that the employer has to exercise a level of judgement, care, prudence, determination, and activity that would be reasonably expected of an entity responsible for a work site and its safety. To exercise due diligence in this regard, an employer must implement a plan to identify possible workplace hazards and carry out the appropriate corrective action to prevent accidents or injuries arising from these hazards. The conditions for establishing due diligence include several criteria.

The employer must have in place a documented safety program that includes policies, practices, and procedures to demonstrate that the employer

- Provides employees with information to enable them to work safely,
- Audits compliance with the safety program requirements,
- Identifies hazardous practices and hazardous conditions,
- Makes necessary changes to correct these conditions, and
- Carries out workplace safety audits.

The employer must provide employee training and education so that employees understand and carry out their work according to the established policies, practices, and procedures.

The employer must train supervisors to ensure they are competent persons as defined in legislation.

The employer must monitor the workplace and ensure that employees are following established policies, practices, and procedures.

The employer must have an accident investigation and reporting system in place incorporating information from these investigations into improved policies, practices, and procedures.

There are obviously many safety requirements of the employer, but workers also have responsibilities. They have a duty to take reasonable care to ensure their own safety and that of their coworkers — this includes following safe work practices and complying with regulations.

---

[2]   **Employer** – The individual or organization that directly pays the employee for work rendered.

III. Documenting the Safety Program

A successful safety program must address site-specific hazards and safety concerns, as well as provide direction to ensure compliance with federal, state, and local safety regulations and standards. Safety program requirements and their emphasis will differ for each project. A project that involves significant work with buried utilities will have very different safety program requirements and emphasis from those of a project with extensive work at heights.

Table 2.1 Considerations when developing a safety program

Employer Considerations when Developing a Safety Program

- Do you understand your safety responsibilities?

- Have you set objectives for safety just as you do for quality and production?

- Have you integrated safety into all aspects of your work?

- Do you have procedures in place to identify and control hazards?

- Are employees encouraged to report unsafe conditions and unsafe practices to their supervisors?

- Have you explained safety responsibilities to all employees and made sure that they understand them?

- Have employees been trained to work safely and use proper protective equipment?

- Do you keep records of the training received by each employee?

- Are managers, supervisors, and workers held accountable for safety, just as they are held accountable for work quality?

- Do your records show the action taken when safety procedures have been knowingly violated?

- Have you committed appropriate resources to safety?

- Is safety a factor when acquiring new equipment or modifying work processes?

- Do you review your safety program at least once a year and make improvements as needed?

A. Management Commitment and Leadership

To promote the integration of safety into regular work practices there must be a clear assertion by the individuals controlling the work that safety is important. Safety begins with the attitude that accidents are preventable and that requirements for safe work practices must be followed. Safety should be included as the first item of discussion on all meeting agendas including those dealing with cost, schedule, sales, quality, etc. Safety programs will only be as good as management's commitment and visible support. The importance of safety over that of expediency must be regularly emphasized so that the workforce understands that although schedule and production are important, work must be performed safely.

1. Safety Program

Management must clearly define its commitment to safety and its expectations of workers in that regard. To document expectations for the benefit of employees, each organization should commit their safety program to writing. This document should define the corporate safety policy and safe operating procedures and it should contain a mechanism to review these procedures and develop new ones as technology and standards change.

BENNETT CONSTRUCTION CO.
BENNETT BUILDING & DEVELOPMENT, INC.
SAFETY POLICY

To all Employees January 1, 1999:

At the Bennett Construction Co. our employees are our most valued resource. Their safety and health are our principal concerns. Management is committed to providing a healthy work environment and preventing the occurrence of injuries.

- *No work is so important that it should be done without due consideration for safety.*

- *Supervisors are responsible for ensuring that their employees are trained in approved safe work procedures to work efficiently, without accidents.*

- *All employees will be held individually accountable for working safely and complying with the relevant laws and regulations incorporated into our safety program.*

Jim Bennett

President

Figure 2.1 Safety policy—example

2. *Policy Statement*

The cornerstone of the safety program is a safety policy. This is a written statement of the principles and general rules embodying the company's commitment to workplace safety and health. The policy statement can be brief, but should mention

- management's commitment to protect employee safety,
- the organization's safety philosophy,
- who is accountable for the occupational health and safety program,
- the general responsibilities of all employees,
- that safety shall not be sacrificed for expediency, and
- before any work is initiated it should be evaluated for potential hazards and consideration given to how identified hazards should be controlled.

The policy should be stated in clear and unambiguous terms. It should convey to the workforce the message that management is sincerely committed to safety and will stand behind it. This is particularly important in an environment where the message is different from one that has been communicated in the past. Each organization must determine how best to communicate its safety policy and rules to its workforce.

One must take the first step to communicate one's commitment to safety sometime. Management must at that time be prepared to deal with the question, "Why is safety so important now when it wasn't before today?" Unfortunately, it often takes a major business disruption or a personal tragedy before organizations make the decision to begin to place greater emphasis on safety. A safety policy statement promulgated by management and followed with specific guidance on how safety is to be incorporated into work practices has to occur for the process to begin.

A major shortcoming in regard to communicating site rules and safety procedures is the failure to regularly reiterate and reinforce them. Managers and supervisors often assume that if the message has been delivered once, that task is done and can be forgotten. How often must teenagers be told to clean their rooms and possibly not get their allowances before it becomes a habit? Although we see speed limit signs on the road regularly, how many times must we see a state trooper issuing traffic tickets or even receive them ourselves before we begin to pay attention to the speed limit on a particular stretch of road? The fact is that only through repetition, regular reinforcement, and leading by example will the message be understood and believed.

Once established, management must be committed to ensuring that the safety policy is implemented without exceptions. The safety policy should have the same importance as the other organization policies.

A contractor's safety rules might look something like this.

BENNETT BUILDING & DEVELOPMENT, INC.
SITE RULES

It is very important to continually earn our customer's business at every stage of construction. Our goal is to provide the best possible product available to these customers. In order to reach this goal, expectations of each and every contractor/person are listed here. There will not be any exceptions to these rules.

1. Job sites must be broom swept at the end of each day and debris placed in a garbage can provided or in the container provided outside only!

2. Smoking is only permitted in designated areas.

3. Employees must wear the personal protective equipment defined in their employer's Job Safety Analysis.

4. Fall protection is required for all work performed above 6 feet.

5. Scaffolding shall be inspected by a competent person before use.

6. Stair and floor hole openings shall all be properly barricaded.

7. No drugs or alcohol are permitted on the job site.

8. Any and all food, packaging, etc. must be removed from the job site each and every day. Do not put in garbage.

9. All lumber that is not in use must be stacked neatly with nails removed.

10. Lock all doors, garage doors, and windows (including basement windows) if last person to leave job site.

Any violations of these rules will result in the following:

1ˢᵗ Violation — verbal warning given and call placed to contractor office.

2ⁿᵈ Violation — will be asked to leave the job site and another person will be expected for replacement.

3ʳᵈ Violation — change of subcontractor.

I appreciate your understanding in this matter.

Jim Bennett

Figure 2.2 Site safety rules—example

3. Safety Rules

Safety rules are needed to simply and concisely define accepted work practices. Poorly defined rules or those rules with little relevance to the work being performed may be interpreted as an indication that safety is not important. Too many rules make compliance with them all difficult. The following are some guidelines for establishing rules.

Rules should be

- available to all employees in written form,
- specific to safety concerns in the workplace,
- stated in positive and understandable terms,
- explained,
- enforceable, and
- reviewed periodically to evaluate their effectiveness and to make changes for improved effectiveness.

One approach to initially establishing safety rules is to address the most frequent causes of injuries on construction sites:

Falls	33%
Struck by	22%
Caught between	18%
Electrocution	17%
Other	10%

If you are really stuck for ideas, determine the most frequently cited OSHA standards (http://whttp://whttp://www.osha.gov/oshstats/) for your particular Standard Industrial Classification (SIC) code category and use those as the basis to begin developing your rules.

Compliance with health and safety rules should be considered a condition of employment. Rules must be explained to new employees when they start work and when they are transferred to new responsibilities. Periodically the rules should be reviewed with employees to ensure they continue to understand those applicable to their work. When establishing rules make sure that you are prepared to apply them all to everyone. That includes even "Bob," your best producer, who sometimes cuts corners. If you are not prepared to apply the rules to him as they are written, then you should reconsider what is important to you and your safety program.

Once it was suggested to me that "You can have as many rules as you want, as long as they fit on an 81/2" by 11" sheet of paper and are double-spaced." That struck me as a reasonable guideline and has been an effective rule of thumb ever since. Once a rule has been totally engrained into the work culture and is no longer an issue, then substitute one that deals specifically with another issue of greater concern.

B. Safe Job Work Procedures

Many job-related injuries occur because employees are not aware of how to control potential hazards related to their work. This may be because they were never trained to perform the work safely in the first place or they may not have received training to help them identify hazards or they do not have the background to develop appropriate controls to deal with those hazards once they have been identified. Think back when you started to work. How did you learn the specifics of performing a manual task? Did a supervisor take you aside and explain how to do the job? Did that person discuss all the details regarding PPE, work steps, and the control of potential hazards? Or, did the supervisor accept your claim that you knew what you were doing and then you figured out on your own what you were supposed to do by watching the people around you?

The obvious hazard of the last situation, and most common scenario, is that the person we were copying may not have learned the safe way of performing the work either. At some point, each work task should be evaluated to identify the most efficient way of performing it safely. The agreed-to-work-safe work practices defined as the result of this evaluation should then be communicated to the individuals required to perform the work and to those responsible for supervising it.

1. Job Safety Analysis

A means of systematically identifying and evaluating safety issues in work is a Job Safety Analysis (JSA). Each step of a task is evaluated to identify the hazards associated with the work. Where issues are identified, an effective means of performing the work in a safe manner is agreed to. This process may involve the participation of several individuals including supervisors, individuals who perform the work, and technical specialists with an awareness of safety-related considerations to help recommend safe procedures to execute the work.

A job safety analysis generally consists of the following steps:

1.   select the job,
2.   break down the job into a sequence of steps,
3.   identify the hazards, and
4.   define preventive measures.

Where a job consists of more than one specific task, each separate task should be analyzed.

Once workers have been briefed on the correct manner in which work is to be performed and the required personal protective equipment, it is up to them to perform the work within defined controls. However, supervisors should monitor their subordinates' work to ensure the procedures have been understood and are being practiced. Supervisors and employees should both provide feedback when the need for corrections is identified in the JSAs.

Written procedures outlining the safe way to perform tasks provide the basis for supervisor job instructions, evaluation of compliance with safe work practices, and evaluations following accidents to determine if safe work practices were followed. Written procedures provide a benchmark against which to judge if changes must be made to accepted work practices to avoid the recurrence of a work-related accident.

A job safety analysis should be developed for all critical tasks with priority given to addressing tasks

- where <u>frequent</u> accidents and injuries occur.
- where <u>severe</u> accidents and injuries occur.
- with a <u>potential for severe</u> injuries.
- which are <u>new</u> or that are a modification to a previous procedure.
- <u>infrequently</u> performed jobs, such as maintenance.

IV.   Responsibility and Accountability

Having made the assertion that safety is a serious consideration, management must then clearly define responsibility for safety. Safety professionals have for years stated that safety is a "line management" responsibility. However, they are often hard pressed to clearly explain what this means and seldom do they address employee responsibility for safety.

For a safety program to be successful, management, supervisors, and workers must recognize that they each have a role in the safety process. Management has a responsibility to provide their employees with a safe and healthful work environment. Management must also ensure that employees have the tools, personal protective equipment, and other resources needed to execute their work safely. Supervisors have the responsibility to ensure that each of their subordinates clearly understands how to perform their work safely and that they follow safe work practices. It is the responsibility of the workers to make sure they understand how to perform the work and have the tools and protective equipment they require to perform the work following approved safe work practices. Safety is not an activity employees must add to their regular work, but rather it is an integral, full-time component of each individual's responsibilities.

Responsibility may be defined as an individual's obligation to carry out assigned duties. Responsibility can be delegated to subordinates, giving them the right to act on the behalf of their superiors. While some responsibilities can be delegated, the superior remains accountable for seeing that assigned duties are carried out. Proportional to this responsibility and associated rewards for performance should be accountability. For a safety program to be effective, individuals must be held accountable for their performance and the results of their decisions in regard to safety. To fulfill their individual safety responsibilities, individuals must know what their responsibilities are, have sufficient authority to carry them out, and have the required knowledge to do so effectively.

Documented safety programs should define the safety responsibilities of individuals at each level of the project organization, from management to the subcontractor's craftsmen. This information will enable everyone on the project to understand who is responsible for what in the project safety program. Since every individual on a job site is ultimately a worker, *worker responsibilities* apply to everyone. First-line supervisors have additional responsibilities and management has duties in addition to those of the worker and supervisor. The following are examples of the types of responsibilities that should be expected of individuals at each project level.

Worker responsibilities include

- knowing and complying with safety regulations;
- following safe work procedures;
- using personal protection equipment as required by the employer and appropriate for the work being performed;
- correcting or reporting unsafe work practices and unsafe conditions;
- helping new employees recognize job site hazards and follow proper work procedures; and
- reporting injuries or illnesses immediately.

First-line supervisors are responsible for

- instructing workers to follow safe work practices;
- ensuring that only authorized, adequately trained workers operate equipment;
- correcting unsafe acts and unsafe conditions;
- enforcing health and safety regulations;
- promoting safety awareness among workers;
- ensuring required PPE is worn by workers and that they understand the reason for its use;
- inspecting their own work areas and taking action to control or eliminate hazards;
- ensuring injuries are treated and reported; and
- investigating all accidents/incidents.

Management responsibilities include

- establishing and maintaining a health and safety program;
- ensuring workers are trained or certified, as required;
- providing workers with safety and health information;
- providing a safe and healthful workplace;
- ensuring personal protective equipment is available;
- supporting supervisors in their safety and health activities;
- providing medical and first aid facilities;
- evaluating safety performance of supervisors; and
- reporting accidents and occupational illnesses to appropriate authorities.

Designated Safety Representative

Although the job site manager is responsible for ensuring that an adequate job site safety program is in place, depending on the project size, it may be necessary to assign a knowledgeable individual the responsibility of overseeing the implementation of the safety program; this is the Designated Safety Representative. Supervisors (including foremen) must understand that they will still be held accountable for their own safety-related decisions and the performance of their respective subordinates. Workers must also be made to understand that they are expected to look out for their own safety and adhere to the safety rules. It is the role of the Designated Safety Representative to bring safety issues and concerns to management's attention and to guide supervisors and workers in the implementation of their respective safety responsibilities. The designated safety representative is not the entire safety program.

An Example of Safety Accountability

Sensitivity to the importance of safety was heightened measurably at one facility following a detailed government safety compliance inspection precipitated by repeated citations for safety violations. It was apparent to the organization's management that they would no longer be in business if they received any further fines for failure to be in compliance with safety regulations. The facility was scheduled for a follow-up compliance inspection in three months. In response to this concern, senior management brought in a new manager to whom responsibility for safety was clearly assigned. This individual understood how to define his expectations with regard to safety and communicate this to his new organization. He explained the situation to the entire organization. He stated that he felt very strongly that not only was work safety important, but that he expected everyone to work safely. He explained that he felt that responsibility for safety was each employee's responsibility and indicated that he would hold each of them individually accountable for their respective performance and results. He made it clear that no violation of safe work practices was acceptable, nor would the failure to correct identified unsafe conditions or work behavior be tolerated. Either situation would be grounds for instant dismissal. The previously stated responsibilities were clearly explained to all employees.

A craftsman was seen shortly thereafter standing on the top of a ladder while his foreman was working nearby. Both individuals were terminated that day. The follow-up safety audit did not identify a single compliance failure, nor was anyone found to be working unsafely. The manager had clearly defined his expectations with regard to responsibility for safety and demonstrated that he was serious about accountability.

A. Identification and Control of Hazards

Once we have ensured that the individuals who have been delegated responsibility for managing personnel and the individual assigned to perform the work understand their respective safety responsibilities, we must ensure that they are able to recognize work hazards.

This is often a learned skill, which through practice becomes an automatic or reflexive process. However, there is such an apparent need for this sort of training that a whole industry has evolved to provide training in this area. E. I. du Pont de Nemours and Company offers their Supervisor Safety Training Observation Program (RSTOP), Argonaut Insurance Co. offers its Supervisor Training in Accident Reduction Techniques (RSTART) program, and Behavioral Science Technology offers their Behavior Based Safety program, to name a few programs. In simple terms, these programs emphasize the fact that work that gets inspected gets done in the manner that those supervising it want it done. If managers, supervisors, and employees are on the lookout for unsafe work practices and behaviors that may lead to an undesirable outcome, then there is an opportunity to correct it. This process of identification of hazards must become ingrained in the work process.

B. Competence Commensurate with Responsibilities

We often mistakenly assume that individuals who have been conferred the status of craftsman, supervisor, or manager, by virtue of the excellence of their technical knowledge, should also be able to recognize safety hazards associated with their work. Our second mistake is to assume that those individuals who are able to recognize hazards will include safety considerations in their work planning, thereby eliminating or controlling the hazards associated with their work. Although some aspects of safety are intuitive, the ability to recognize hazards and to develop means by which to control the hazards must be learned by most individuals.

We must make sure that the supervisors we have delegated to assign work understand their safety responsibility and have the technical background to fulfill this responsibility. We must also make sure that they make similar assurances regarding the training and safety awareness of the individuals to whom they, in turn, assign work. The apparent failure of individuals with supervisory responsibility in construction and general industry to fulfill this responsibility has resulted in the establishment of a new work classification, the "Competent Person".

Competent Person

A Competent Person is generally defined as an individual, who as the result of training and/or experience, is capable of

- identifying or predicting hazardous situations,
- authorizing prompt corrective measures to eliminate them, and
- doing so.

The fact that regulators have felt the need to define a requirement for a *Competent Person* classification suggests that in many cases individuals typically assigned supervisory and oversight responsibility for work have not been effectively identifying and controlling known hazards. Competence is about action not certification. See Table 2.2 for a listing of some of the Competent Person categories and some responsibilities identified in 29 CFR 1926 and various ANSI Standards.

V.   Reinforcement (Discipline)

Consequences control behavior and behavior is based on work direction and work conditions. A mechanism must be in place to address situations where the violations of safety principles are identified. It is the accepted practice in construction that once training, guidance, and encouragement have been exhausted, discipline is the last remaining recourse to reinforce the application of safe work practices.

Disciplinary programs in contracted work environments generally follow the same framework. Individuals identified as violating safety rules are first given a verbal warning. A second violation results in a written warning and possibly being barred from the site for two days. (The host employer cannot generally fire a contractor's employee, but can deny the individual access to the job site.) The third time the individual violates a safety rule, the employee is barred from the site for six months or denied access to the site for the duration of the project.

When an employee is involved in a case of <u>serious or gross misconduct,</u> progressive disciplinary action is not generally invoked. An immediate and thorough investigation of the incident is conducted, and penalties may include termination for cause or denial of site access for the remaining project duration.

Table 2.2 Competent person categories

Excavation	Underground Construction
Excavation Design	Evaluation of Equipment
Structural Ramps	Haulage Equipment
Water Removal Eqpt.	Hoisting Equipment
Soil Conditions	Drilling Equipment Underground Utilities
Hazardous Atmospheres	Work Environment
Shoring Requirements	Ventilation
Scaffolds	Air Monitoring
Evaluation of Equipment	Compressed Air
Structural Integrity	Geological Evaluation
Scaffold Components	Soil and Rock Conditions
Connections	Personal Protective Eqpt.
Wire Suspension Ropes	Engineering Controls
Fall Protection	
Tie Points	*Other areas where "Competent Persons"*
Safe Access	*are expected to be designated are where*
Modifications/Alterations	*there is present:*
Selection of Assemblers	
Work Conditions	**Ionizing Radiation**
Fall Protection	**Flammable Liquids**
Certify Nets	**Painting**
Personal Fall Arrest Systems	**Lead**
Monitor Safety of Employees	**Slings**
Implement Fall Protection Plan	**Ladders**
Cranes & Derricks	**Material Handling**
All Machinery Prior to Each Use	**Hearing Protection**
Annual Inspections	**Powder Activated Devices**
Equipment Following Trial Lifts	**Helicopters**
Hazardous Chemicals	**Respiratory Protection**
Identification of Their Presence	**Shipyards**
Establishing Regulated Areas	**Telecommunication Systems**
Exposure Monitoring	**Blasting Agents**

Nowhere in this practice does the employer responsible for the work being performed unsafely assume any blame. In an environment where the presence of such a disciplinary program is in place, construction crews generally keep quiet about unsafe acts and will only disclose recordable and lost time accidents if it cannot be avoided. As it is our objective to establish a safe work environment free of recognized hazards, it is in our interest to better understand the factors leading to unsafe behaviors. Individuals do not generally come to work with the objective of injuring themselves.

Note   *Do not resort to implementing this process until the reason for the unsafe behavior has been clearly established to have been a wanton safety violation. Only then, in circumstances where the worker must be permanently denied access to the job site or the contractor's contract must be terminated for serious or gross unsafe behavior, should you consider imposing this penalty.*

Safety violations and injuries occur because of three factors: personal factors, work factors, and those factors out of the worker's control.

Personal factors include the lack of skill or knowledge to perform the work properly. A personal factor may be lack of motivation to perform the work properly or the perception that doing the work according to approved procedures is not important. Perhaps the individual performing the work believes that since doing it the correct way takes more time and effort than the shortcut he has been using with effective results, it is fine as long as no one has instructed him to do the work any differently.

Work factors include inadequate work procedures or standards that direct the safe way of doing work. Safety violations or injuries may have been the result of lack of clear communication of expectations regarding implementation of approved procedures and work practices. It could be that the tools and equipment provided to perform the work were not adequate.

The last set of factors is those conditions over which the employee had no control, such as job site hazards of which the worker was not aware or those created by another contractor that were not communicated.

When an accident does occur, focus on identifying what must be changed in terms of work procedures, equipment, and educating the workers. Punishment only motivates workers to remain secretive about unsafe acts that could become the source of a serious accident. Is the power supply for a device that must be locked out or reset located in a remote location the reason why workers failed to lock out the breaker feeding power to a piece of equipment? Is the device failing frequently, requiring that it be reset regularly the reason that the safety alarm has been disabled? Although logically management would suggest that they be advised of such circumstances or that the employees take the time to follow safe work procedures, this may not be a practical answer. Perhaps workers have reported the problems before, in fact, frequently. If they didn't see any change in circumstances, they may have drawn the conclusion that it isn't important to management and that their only recourse is to work on the device without locking it out. Otherwise they would be wasting time going back and forth between the breaker and the device to lock it out "properly."

Occasionally you will meet a superintendent or project manager who will not be responsive to this proactive approach. The host employer must then replace this individual with one who better understands the value of developing a responsive and well-trained workforce.

VI.   Inspections

Regular evaluation of job sites must be conducted to verify that workplace hazards are being controlled and that safe work practices are being followed when work is in progress. Traditionally, safety inspectors and occasionally, supervisors conduct safety inspections. This approach, as we are learning, is limited in value as it is restricted to the availability and technical knowledge of a few individuals. In a progressive environment, management and supervisory personnel as well as other project team members and workers should participate in safety inspections.

Each organizational tier in a project hierarchy, from the client to the subcontractor performing the work, has a vested interest in ensuring that safety is integrated into work practices. Each organizational element should conduct inspections that focus on issues of concern at their respective levels. For example, the client should verify that the construction manager is enforcing a requirement that the general contractors have safety programs, and so on down the chain of command.

Field engineers should inspect their work areas prior to the start of work and on a frequent basis, to identify existing or potential hazards not anticipated and planned for during the project hazard analysis or covered in the safety program. They should monitor job-site activities under their control to assess the effectiveness of their safety plan implementation and to identify hazardous conditions or activities that need immediate correction or require additional training or retraining.

Since the greatest risk of injury is present at the worker level, each contractor should be required to conduct and submit regular safety inspection reports of their own work to the job-site manager or the host employer's field engineer. This compels the contractor to participate in the process of looking out for conditions and work practices that could put their own employees at risk of injury. Contractor inspection reports also serve to document the contractor's involvement in the identification and correction of unsafe work conditions and work practices.

Current safety legislation requires workplace inspections be conducted on a "regular basis" to ensure workplace safety. The frequency of planned formal inspections must be defined by each organization based on the rate of change of physical conditions on the job site, their record of compliance with safe work practices, and the frequency of accidents and injuries. Individuals developing safety programs often overlook the fact that field supervisors conduct inspections on a regular basis as part of their routine work. They are looking for defects in quality, opportunities to improve productivity, and symptoms of construction problems. Safety should be included in this consideration. Participation in formal inspections should be part of every supervisor's and manager's job description.

The safety program should define

- who should conduct inspections,
- when inspections should be conducted,
- the focus of the inspections,
- who should receive the reports,

- how deficiencies should be addressed, and
- what records should be kept.

VII. Training

Training should be an integral part of all safety programs. Only through regular communication of hazards, control measures, and safe work practices can the incidence rate of work injuries be systematically reduced. The most frequent accident root cause and the OSHA citation most frequently issued following workplace accident investigations are related to deficiencies in employer training programs. To perform their work safely, workers must be trained. For the training to be effective each job must be thoroughly analyzed and safe work practices developed and communicated to the workforce.

According to the Business Round Table, approximately 250,000 new workers join the construction workforce each year. About 80% are non-union and do not have the benefit of formal construction trades training or safety training for that matter. This is a significant concern given the increasing sophistication and complexity in construction. The unfortunate result is the very high fatality and injury rate being experienced by construction workers. In 1995, 1,033 construction workers died from on-the-job accidents. 930 died as the result of worker errors. This is equal to 9 or 10 airline crashes a year. Just as traumatic are the multitude of disabling injuries and the thousands of lost work-time accidents that occur for each fatality. There must be a process in place to ensure the individuals coming to the job site are aware of project-specific safety issues and are trained to handle them.

As a point of reference, between 29 CFR 1910 and 29 CFR 1926 there are 192 non-overlapping requirements for the provision of training. Clearly, a single employer will not perform all the activities that require training. The fact remains that there are a significant number of requirements for training and employers should be aware of this and have a process in place to address the needs of the workers. OSHA has placed the burden on employers to provide their employees with safety training. There is no guidance regarding the frequency of training, but information that was learned just a few years ago regarding the use of safety belts, for example, is no longer applicable given current requirements for body harnesses for fall protection.

The host employer should ensure that everyone who comes to their facility be familiar with the site-specific safety issues. Since it is a fact that few employers are in a position to implement trades training programs for their employees, the host employer should require an assertion from each contractor that each employee understands the hazards associated with their work and is able to perform their work safely. The host employer might define an expectation that each employee should be briefed on the work to be performed that day and be given a means to address specific safety issues that may surface as the work evolves. Safety training and safety communication mechanisms such as the following are effective vehicles.

31

A. Safety Orientations

Job-site safety orientations provide a forum for the host employer to convey their commitment to providing a safe working environment. The orientation is an opportunity to point out site-specific information regarding the facility's safety hazards and it is an opportunity to remind workers of their responsibility to give due consideration to safety while conducting their work. Even the most seasoned workers and supervisors need to become familiar with job-site layouts, the project management personnel, company policies and other information related to an unfamiliar project. This is the time to explain what is expected of them should they perform work that exposes them to a fall hazard, released toxic fumes, excessive dust, etc. It is an opportunity to remind workers to bring hazards they might create to the attention of their supervisors so that those individuals who might be affected by them can be advised.

Sadly, safety orientations are often the exception rather than the norm on most job sites. This step in the construction sequence is often skipped in the interest of the expediency of getting the workers onto the job site to start the work. Safety awareness and hazard communication training should be conducted before any individual is permitted to begin work on or visit an unfamiliar site. Statistically, the majority of injuries occur to employees who are not familiar with site-specific safety expectations and those individuals who are not familiar with the job-related hazards. Approximately 50% of construction industry deaths occur to individuals who have been on a job site less than 30 days. Accident frequency decreases with increased experience and greater safety awareness. The awareness of safety issues can be accelerated through safety and health training.

Orientations must be conducted with sufficient frequency to enable contractors to get their personnel trained and onto the job site, otherwise they may find a way to circumvent the process. The content of the orientation should be tailored to the needs of the individuals being briefed. Visitors and contractors delivering material do not necessarily need the same course detail required by a specialty contractor.

The contents of site orientations should include as a minimum:

- introduction to the job site,
- site rules and regulations,
- HAZCOM information,
- requirements for personal protective equipment,
- fire protection system and emergency procedures,
- first aid and treatment program,
- immediate reporting of injuries,
- permit requirements,
- disciplinary program, and
- introduction to key project personnel.

The bottom line that must be regularly reinforced is the fact that working safely is the corporate objective. No one has authority or permission to work unsafely or to take shortcuts that place him or her or the project at risk.

One means of ensuring every person on the job site has attended an orientation is to issue hard hat decals on completion of the training. Another means is to issue site-access picture badges to avoid attempts to bypass the orientation requirement. Issuing identification badges is not a complex or costly process given the technology available today.

A brochure with a concise summary of the project safety program and site rules should be provided to each orientation attendee. The brochure should be a reminder of the important points that were addressed during the orientation. Generally a single safety training session is not sufficient to convey all the requisite information needed in construction and to maintain a heightened sense of safety awareness. To reinforce what has been addressed at the site safety orientation, regular follow-up training is required. As we have a job to do, we must have a seamless way of integrating safety into our routine work processes. The first opportunity is the pre-work meeting.

B. Pre-Work Meeting

The most effective time to reinforce safety is at the start of each workday when each supervisor assembles his work crew to delegate that day's activities. Each supervisor should take this opportunity to reinforce the importance of safety and associate it with the work to be performed.

Pre-work meetings held at the start of each day should be short and include the following agenda:

- Review previous day's accomplishments
  - What was done well
  - Opportunities for improvement
- New work assignments
  - Work practices (JSA)
  - Safety issues
- Questions and answers

Work practices will improve to the extent individuals are provided with feedback on what is expected of them. A review of the previous day's activities with an emphasis on lessons learned and opportunities for improvement provides a forum for continuous improvement.

New work assignments and a review of the associated JSA for the work to be performed that day are then addressed. During this discussion the supervisor confirms that everyone understands what is expected of them. The foreman should solicit feedback from the workers on the content of the JSA and what changes if any might be required to complete the work safely. To the extent workers have been on other jobs they will most likely be able to contribute valuable suggestions to improve the effectiveness of the JSA. Once they have critiqued it and have received positive feedback from the supervisor regarding their observations, they will be more likely to be committed to following the JSA.

The pre-work meeting also provides an opportunity for the supervisor to speak with each employee to determine if they are prepared to work that day. The supervisor has the opportunity to evaluate each worker's behavior and to judge if alcohol, drugs, or some preoccupation they may have brought from home could impair their ability to perform their work effectively. Workers who are unfit for work are a danger to themselves and to the rest of their work crew.

The supervisor should document the main points addressed at pre-work meeting in his daily log as a record of what was discussed for future reference.

C.   Onsite Safety Coordination

If anything can go wrong, it will. Consider that work is in full swing and a load of sheet rock has arrived. The location where the material was to be unloaded is blocked. The individual delivering the material proposes unloading it over vehicles nearby to be able to go on to his next delivery. The supervisor is not immediately available. What happens now?

Employees should be encouraged to bring safety issues whose solution is not clearly evident to their supervisor's attention. They should be made to understand that it is the right thing to call on the supervisor or even the job boss if necessary. In today's complex work environment we cannot afford to allow employees not to call, nor can we afford for them to assume personal risk for the sake of getting a job done quickly. "The hand rails are missing in this area, I won't get too close to the edge." "We don't have a scaffold, I can borrow a ladder and work on it to get the job done." Workers may, out of a misguided sense of dedication, opt to perform work that places them at risk of personal injury or damages other project equipment. Once a worker is hurt, the supervisor's time, the worker's time, and so on will be absorbed for periods far in excess of what it would have taken to have discussed a safety issue and resolved it. Make it clear that any employee who identifies a safety concern may call a meeting of the individuals immediately involved in the work and the supervisor to resolve the safety concern.

Employers should encourage a process at the pre-work meetings that fosters positive feedback for employees that bring forward safety issues or concerns. Employers should encourage employees to be on the lookout for opportunities to improve the safety of their work.

D.   Weekly Toolbox Talks

Traditionally, contractors who remain on a job site for two or more weeks are required to conduct weekly *Toolbox Talks*. The toolbox talks are expected to last no less than five minutes and to address topics relevant to the work being performed.

Weekly toolbox talks provide the opportunity to reinforce the company's safety policy, changes in safety procedures, to review lessons learned from accidents that may have occurred within the company at other sites, to address safety topics of a general nature, or to discuss the development of JSAs for upcoming work.

The downside of toolbox talks is that it is not reasonable for supervisors to be able to anticipate all the safety issues that will surface during the next week, nor is it reasonable to expect the workers to retain this information or put it into practice several days later. It is a common practice among some companies to send purchased toolbox talks to the field that may have nothing to do with the work being performed. For this reason, pre-work meetings and onsite safety coordination are powerful safety communication tools.

Weekly toolbox talks are a good forum to review general safety issues and to summarize lessons learned over the period of the previous week. They are also an opportunity to train workers in new developments in safety subjects.

VIII. Accident Reporting and Investigations

Occupational safety and health legislation requires that all recordable personal injuries be reported. There may be minimum legal requirements triggering these investigations, but realizing the value in conducting investigations, many organizations also investigate lesser accidents where damage did not involve injuries and were only "near misses."

Accidents and incidents are investigated so that measures can be identified to prevent a recurrence of similar events. Although investigations represent an "after-the-fact" response, a thorough investigation may uncover hazards or problems that can be eliminated to prevent the occurrence of future incidents.

The safety program should specify

- what is to be reported,
- to whom it is to be reported,
- how it is reported,
- which incidents are investigated,
- who will investigate them,
- what forms are used,
- what training investigators will receive,
- what records are to be kept,
- what summaries and statistics are to be developed, and
- how often reports are to be prepared.

IX.   Housekeeping

Slips, trips, and falls are the leading cause of accidents on construction work sites. Litter and debris conceal tripping hazards and increase the possibility of other injuries. As a general rule, a site with poor housekeeping practices also has a poor safety record. Debris on a site also creates extra work, as it frequently needs to be relocated for work site access. However, as with any other objective, there must be regular reinforcement of the importance of this requirement.

Nothing sends a clearer psychological message to employees and subcontractors than management's commitment to providing a safe and healthful work environment than their insistence on an orderly project work area.

X.   Substance Abuse Program

Consider that eleven percent (11%) of the drivers on U.S. roads today do not have a valid driver's license. These individuals account for 90% of all traffic accidents. Then consider the fact that the Center for the Protection of Workers' Rights has determined that in some regions of the U.S. up to 30% of construction workers report to work under the influence of some behavior modifying substance. One cannot help but wonder how many human error accidents are attributable to substance abuser error. The use of behavior-altering substances affects job performance and also places the worker and his peers at risk. Any worker, regardless of ethnic origin, socioeconomic background, or occupation could be a substance abuser and it is not always the substance abuser who is injured. Drug counselors estimate that a chronic substance abuser functions at between 50 and 67% of his capacity on the job.

Substance abuse is widespread throughout the United Sates. Unless an organization institutes a substance abuse program, drug and alcohol abusers probably work there. Therefore, substance abuse screening is becoming a necessity in today's workplace. Companies that screen job applicants report that initially a high percentage of candidates test positively and as word spreads of the company's requirements, this number reduces to between 5 and 8% of applicants.

Drugs and alcohol

- Cause workers to take more chances,
- Increase the potential for injury,
- Increase absenteeism and tardiness,
- Decrease productivity and quality of workmanship,
- Increase health insurance and workers compensation costs, and
- Increase theft of materials, tools, and equipment.

A. Workplace Drug Testing

To identify employees who abuse drugs, a workplace drug testing program must be in place. This requirement is generally opposed in construction unless it is for cause. However, approximately 20% of the American workforce has a drug testing policy in their workplaces. Companies that do impose drug testing requirements send a strong message that they support a drug-free environment. Surprisingly, most employees approve of drug testing programs.

There are several approaches to workplace testing:

- Pre-employment,
- Post-accident or for cause testing,
- Scheduled testing (during employee medicals), and
- Random testing.

It is imperative that organizations implementing a drug-free workplace program have written policies and procedures in place. A lawyer experienced in labor and contract matters should review these and help design a program that complies with state laws and meets the specific needs of the employer.

In the contracted work environment, the host employer should establish drug-testing requirements in the contract language. At high-risk facilities such as refineries and chemical plants, the constructor or service provider should be required to provide documented evidence that all employees have been tested. The contract may further stipulate that testing for cause will be the host employer's prerogative.

A study of the economic impact of substance abuse treatment found these significant improvements in job-related performance:

- 91% decrease in absenteeism,
- 88% decrease in problems with supervisors,
- 93% decrease in mistakes at work, and
- 97% decrease in on-the-job injuries.

XI.   Emergency Procedures

Emergency procedures are plans for dealing with emergencies such as major injuries, fires, explosions, major releases of hazardous materials, violent occurrences, or natural hazards. When such events occur, the urgent need for rapid decisions in a short time can lead to chaos if there is a lack of resources and trained personnel to deal with the situation. At a minimum, procedures for prompt medical response must be established. OSHA expects that any employee should receive medical response within 3 to 4 minutes. This requires that at a minimum there must be someone on each work crew or shift with First Aid and CPR training.

The objective of the plan is to prevent or minimize fatalities, injuries, and damage. The organization and procedures for handling these sudden and unexpected situations must be clearly defined. The process of establishing an emergency plan is not all that complex and is strongly advocated by individuals and organizations that have been caught without one. However, few organizations have established emergency procedures or a crisis management plan.

To develop a set of emergency procedures, compile a list of the hazards, for example, fires, explosions, earthquakes, and floods. Identify the possible major consequences of each: casualties, damage to equipment, or impact on the public. Determine the required countermeasures which might include evacuation, rescue, fire fighting, etc. Inventory the resources needed to carry out the planned actions: medical supplies, rescue equipment, training personnel. Based on these considerations, establish the necessary emergency organization and procedures. Communication, training, and periodic drills are required to ensure adequate performance when the plan must be implemented.

XII.   Conclusion

To a great extent, the success of an effective safety program depends upon management and their involvement in the program. Although one cannot place a dollar value on the humanitarian aspects of a safety program, it is also impossible to place a dollar value on the negative effects personal injuries and fatalities have on labor relations and publicity. Merely adopting a safety program will not yield the desired results without a serious and persistent management commitment to make the program work. It is human nature to place emphasis on that by which one will be evaluated. Thus, safety will receive attention proportional to the importance placed on it by management. Each job site should have a documented safety program in place that addresses the issues and concerns related to site-specific safety hazards.

Since owners ultimately pay for losses on job sites, it is good management practice to oversee subcontractor activities to ensure that they, too, are applying good management and safe work practices. Fear of incurring liability for becoming involved should not be a reason for failing to take an active role in defining the requirements of an effective construction safety program. Owners and contractor managers may well incur greater liability if they do not become involved.

Chapter 3

Safety in the
Construction Process

Successful construction projects are the result of effective planning, execution, and the collective effort of the entire construction project team. However, only when project teams integrate safety into their planning and execution of routine work practices will the success of the safety program be assured as well.

I.   Introduction

Defined in this chapter are the sequence of events common to most construction projects and the associated safety responsibilities of the individuals involved with this process. The duration of the events contained in this process will vary depending on the complexity of the work, technical competence of the individuals involved, and their familiarity with the process. For example, the duration of the planning phase preceding the installation of a fan unit will be quite different than it will be for the construction of an office complex. However, the steps defined here, when followed, result in work projects with very few safety problems.

II. Engineering and Design Phase (Step I)

A. Conceptualize Project

> **Originator/Client's safety responsibilities during the design and planning stage:**
>
> - Be satisfied that only competent people are appointed as coordinators, designers, and constructors.
> - Ensure sufficient resources, including time, are allocated to enable project to be carried out safely.

All work begins with an individual, the *Originator*, who identifies the need for a new installation or modification to an existing facility. The originator's concept is generally presented to an architect or engineering design group (designers) to be developed. This is the case since most originators lack the technical background to convert their concept into a format with the detail necessary to guide a constructor to complete the work as conceptualized.

The originator's contribution to safety is the identification of safety and health issues that may not be readily apparent to the designers. The originator may be aware of environmental issues such as subsurface contamination or of the fact that there are sporadic, very high, ambient noise levels in the vicinity where the work is to be constructed.

B. Identification of a Coordinating Engineer

An architect or engineer is appointed to coordinate the development of the project design. Each organization undoubtedly has their nomenclature to identify this individual. For the sake of simplicity, we will refer to this individual as the *Coordinating Engineer*. The role of the coordinating engineer is to facilitate development of the project design by ensuring that

> **Coordinating Engineer's safety responsibilities during the design and planning stage:**
>
> - Coordinate the identification of safety and health considerations during the design and planning phase of the project.
> - Initiate the ESH Baseline review.
> - Begin the development of the safety and health program and safety file.
> - Coordinate the 90% design review.
> - Ensure that the safety and health file for the work/project is established.

relevant project planning information is communicated to the designers and that they take proper account of safety and health considerations in their design work.

C. Initial Cost and Schedule Estimate, and Baseline Safety Review

The *Coordinating Engineer* should first develop a preliminary cost and schedule estimate for the work as it is understood, with consideration of the safety issues associated with the work, and present this information to the originator.

Originators often have an unrealistic concept of the time it takes to develop a project design and complete the work. They also often have an unrealistically low appreciation of the cost of construction. Although not immediately apparent, both of these aspects of projects have significant safety and health implications.

Unrealistic schedule and cost expectations may induce the originator to attempt to suggest shortcuts to the work process when faced with schedule and cost over-runs. Therefore, the originator must be given a fair initial cost estimate to permit that individual to determine if adequate resources are available to fund the project or if the size of the work project must re-scoped. Potential issues such as schedule delays should be addressed with the originator to determine if special expediting will be required to meet any particular project's completion requirements.

Baseline Safety Review

Prior to the start of the project, the coordinating engineer must conduct a baseline safety review to identify as many of the safety hazards as possible that are associated with use of the space in which the project will be built, and considerations of the subsequent occupancy and maintenance requirements of the completed work. The baseline safety review will assist the designers identify pre-existing and potential hazards associated with the construction and operation of the completed facility. The following specialists may be included in performing the baseline safety review:

Current Occupants
Safety Engineering
Fire Protection Engineering /Fire Departments
Industrial Hygiene
Environmental Compliance
Emergency Management

Armed with the information compiled in the baseline safety review, the designers can identify the means to mitigate or control known hazards during the engineering of the work. This will help provide for a safe work environment for the constructors and avoid the cost of developing controls for these hazards during construction. It will also avoid the greater cost of retrofitting hazard controls for the benefit of the facility operators following the completion of the work.

D. Design & Engineering

The Designer's safety responsibilities during the design and planning stage:

- Structures are designed, to the extent possible, to avoid or minimize risks to health and safety while they are being built and maintained.

- Where hazards cannot be engineered out, provide adequate notice to the constructor.

The designers define the configuration and components of the work through the plans and specifications. The nature of the design influences the means and methods of the project construction. Therefore, designers play a critical role in the identification and control of safety hazards that may exist on the jobsite or may arise as the result of the facility's construction and ultimate design. Many hazards faced by the construction workers are created by the designers, as are the hazards faced by maintenance and operations personnel of the completed work. Figure 3-1 provides a graphical representation of the steps in the engineering and design phase of the work/project development.

E. 90% Design and Constructability Review

When a project's design is approximately 90% complete, an evaluation of the design should be conducted to ensure the environmental, safety, and health concerns identified in the *Baseline Safety Review* have been adequately addressed. Participants in this review should be representatives from organizations with a vested interest in the successful outcome of the project. These should be individuals expected to oversee the contracted work and occupy, operate, and maintain the completed project.

The types of questions which should be considered in this review should include the following:

- Can the facility, as designed, be constructed safely?

- Does the design meet the needs of the occupant and NFPA Life Safety Code requirements?

- Will maintenance and operations personnel responsible for keeping the facility operating be able to do so without personal risk?

The answer to the first question begins with a systematic evaluation of the probable work sequence each contractor will most likely follow, from mobilization until the work is complete.

Comments and suggestions generated by this review enable the designers to re-evaluate the project plans to determine where additional safeguards must be considered. With access to this information the job-site manager may be able to eliminate or minimize hazards through judicious site layout and when planning work sequencing. Where identified hazards cannot be reasonably controlled,

> **Facility Safety Representative's key safety responsibilities during the design and planning stage:**
>
> - Give advice, as requested, to the coordinating engineer on the adequacy of provisions for safety and health features in the design.
> - Assist with the conduct of the ESH Baseline review.
> - Coordinate involvement of safety representatives during the 90% design review.

information regarding these hazards can be brought to the constructor's attention at the Pre-Bid meeting through use of the contract's language.

It is unfortunately often the case that in the interest of expediency the 90% Design Review is not conducted. The risk of by-passing this step is the potential for delays after construction has started when the originator or an inspector identifies needed changes that could have been addressed during the review of the plans. Additionally, safety considerations not addressed during the initial design phase often cost significantly more to retrofit following the completion of the project.

Representation in the 90% design review for safety should include

- Originator
- Maintenance
- Safety
- Scheduler planning
- Estimators
- Procurement

The design review is a good opportunity to introduce procurement and the field engineer who will be assigned to oversee the construction of the project. Having the opportunity to participate in discussions associated with the project design will help them develop an insight into issues that should be addressed in the procurement documentation and subsequently during work planning.

F. Pre-Work Planning

While the design is in progress the coordinating engineer and the field engineer, who will manage the contracted work, have the opportunity to evaluate the anticipated means and methods required to complete the work. Through this evaluation process and with the results of the previously completed baseline safety review, they will be able to develop an inventory of potential hazards. Those hazards which cannot be engineered out of the project or minimized through judicious work planning must be brought to the attention of the bidders during the pre-bid meeting and included in the contract language to minimize the potential problems that may be overlooked. Information regarding the hazards

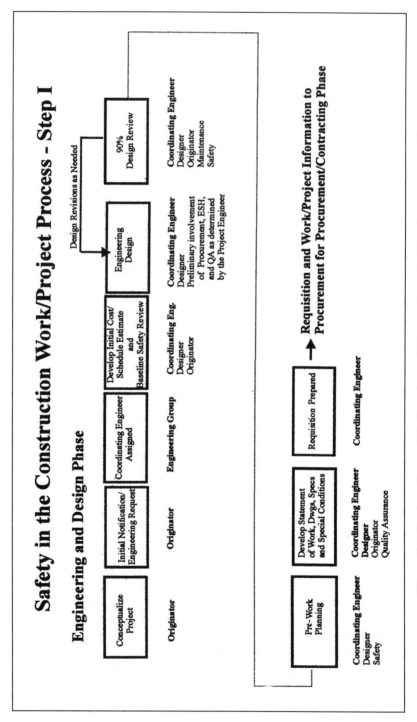

Figure 3.1 Engineering and design phase of the work project development – step I

should also be documented at this point for the benefit of the project or job-site manager so that it can be flagged in the project schedule.

G. Statement of Work

Potential hazards associated with the work to be performed should be included in the statement of work along with the description of the work to be performed and the list of special conditions. At this time the role of the Coordinating Engineer draws to a close and the work is assigned to a Field Engineer who will be responsible for managing the execution of the work.

III. Procurement/Contracting Phase (Step II)

A. Bid Package

The procurement group assembles the bid package that consists of the scope of work, construction drawings, and specifications. Detailed reference should be made in these documents to specific safety hazards that will be the responsibility of the successful bidder to control. The steps in the procurement and contracting phase are outlined in graphical form in Figure 3.2.

> **Host Employer's key safety responsibilities during the Procurement/Contracting Phase:**
>
> - Consider safety and health issues when preparing and presenting invitations to bid.
> - Develop the safety and health program for the project.
> - Verify that contractors selected to perform work have an appropriate safety and health program.

B. Invitation to Bid

Procurement issues an invitation to bid to prospective contractors and sets a date for a pre-bid meeting.

C. Pre-bid Meeting

The pre-bid meeting provides the forum to advise prospective bidders of the scope of work to be considered and to emphasize the importance placed on safety, along with QA and other criteria specific to the work. It also offers the opportunity for contractors to raise questions and clarify any issues they might have with regard to the work and associated safety requirements.

D. Bid Review, Evaluation, and Contractor Selection

On receipt of the bids a technical safety evaluation of the apparent low bidder's ability to perform the work safely is completed. This evaluation is to confirm that the low bidder is able perform the work safely as judged by the technical selection criteria. (If contractors have been prescreened, this process would have been completed prior to the Invitation to Bid.)

45

E. Contract Award

The lowest bidder who meets the technical selection criteria is notified of their selection by procurement. The chosen contractor is asked at this time to produce their project-specific Safety Program and Job Safety Analysis for the work to be performed. The contractor is also asked to identify to whom they have delegated the specific responsibility of coordinating safety-related issues and to provide a list of their designated competent persons, as required.

F. Pre-construction Meeting

> **Field Engineer's key safety tasks during the procurement and contracting phase:**
>
> • Bring to the attention of the prospective bidders information regarding significant safety and health hazards associated with the work/project and the risks of other work that will be underway in the area of the work/project.
>
> • Develop responses to contractor queries via formal addenda to the invitation for bids.

Once the successful contractor is selected, a meeting should be held with the supervisory personnel who are expected to perform the work depending on the contractor's familiarity with the facility and where appropriate, facility personnel should be included in this meeting. The objective of this meeting is to establish that the contractor has a clear understanding of the contract scope, as well as the job-specific hazards and the job's safety requirements. The meeting should review how the contractor will conduct business and the procedures the contractor and other facility organizations will follow before the start and during the performance of the work.

G. Safety Program and Job Safety Analysis Review

The Field Engineer reviews the contractor's safety program and job safety analysis for completeness and applicability to the work to be performed. The objective of this review is to ensure the contractor understands the risks associated with the work and has proposed reasonable safeguards. Where the need for changes to their safety controls are identified, these are brought to the contractor's attention for resolution. Once all documentation is in order a pre-work release meeting is scheduled.

H. Pre-Work Release Meeting

The field engineer's safety expectations are reviewed with the contractor's supervisors. Often this is the first time they may have heard the project's safety requirements. It is not uncommon that the individuals who received the request for bid and attended the pre-bid meeting do not communicate this information to the field personnel. Once all requirements for documentation have been satisfied, and it is clear that the contractor's field supervisors understand the project/facility safety requirements, the contractor is issued a notice to proceed by procurement personnel.

> **Facility Safety Representative's responsibilities during the procurement and contracting phase:**
>
> - Provide guidance for the selection of contractors.
> - Assist the field engineers and coordinators as they evaluate the hazard risks in the work to be performed.

IV. Work/Construction Phase (Step III)

A. General Worksite & Hazard Communication Briefing

When the constructor's employees initially report to work at the jobsite, they should be briefed on the site rules and regulations, the general hazards and special emergency response requirements specific to that work site before they begin to work. Workers need to be aware of the hazards peculiar to the site. Increasingly, Site Hosts are communicating this information to contractors themselves to ensure it is done to their satisfaction. Refer to the steps defined in the work/construction phase outline shown in Figure 3.3.

> **Contractor's key safety responsibilities during the Work/Construction Phase:**
>
> - Provide relevant information on safety and health risks created by their work and how they will be controlled.
> - Develop and implement a site-specific safety and health program.
> - Ensure their workers are following safe work practices.
> - Bring hazards that might affect others to their attention or the attention of the field engineer.

B. Specific Work Site Safety Briefing

Where appropriate, contractors should be briefed on facility-specific work practices and requirements such as Lockout/Tagout or work entry permit requirements. The workers should be informed at this time of the specific individuals who will approve permitted work and how they are to be contacted. They should be briefed on all site-specific safety hazards and applicable emergency response requirements.

C. Job-Specific Training

Contractors should review the Job Safety Analysis (JSA) they prepared for their specific job with their respective employees prior to the start of work. Contractor employees should sign the JSA indicating they have read or been briefed on the information contained in the JSA and understand it. The JSA should be available for the field engineer's review and for the reference of the workers.

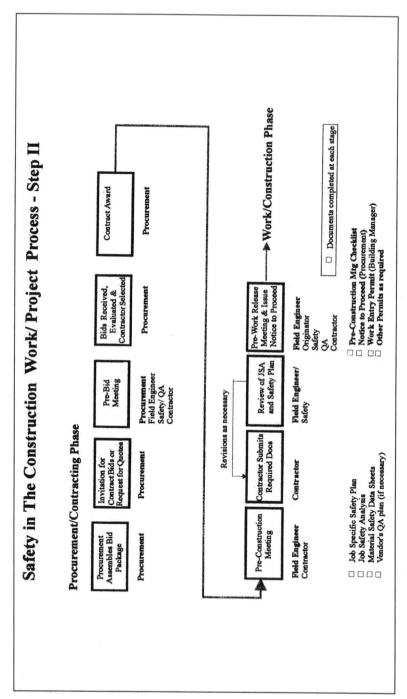

Figure 3.2 Procurement and contracting phase of work/project development – step II

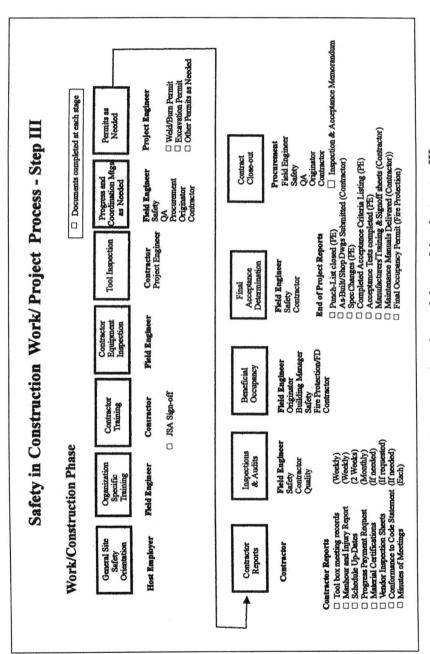

Figure 3.3 Work construction phase of the work/project – step III

D. Contractor Equipment Inspection

All heavy equipment, such as manlifts and cranes, should be inspected for general condition and serviceability prior to being permitted onto the job site. The inspection should ascertain that the equipment appears to be in good operating condition, that appropriate safeguards are mounted on the equipment and that the operators are certified to operate the equipment.

E. Contractor Tool Inspection

Field Engineer's key safety responsibilities during the work phase:

- Be satisfied that contractors are following recognized safe work practices.
- Ensure required training for safety and health is completed.
- Allow only safe equipment and authorized people into construction areas.
- Assist contractor in obtaining safe work permits where necessary.

Prior to starting work contractors should affirm that the tools and equipment to be used on site are serviceable. Tools include all manner of material being used by the contractor such as extension cords, hand and power tools, etc. The field engineer and facility safety representative should periodically assess the condition of tools in use by the contractors as they perform their routine site evaluations. This is one measure of the contractor's implementation of the safety program.

F. Progress and Coordination Meetings

Regularly scheduled project coordination meetings should include safety as the first order of business.

G. Permits as Needed

The object of permits is to ensure that work known to contain recognized hazards is performed following accepted safe work practices.

Facility Safety Representative's key safety responsibilities during the work phase:

- Be reasonably satisfied that contractors are following recognized safe work practices.
- Ensure required training for safety and health is completed.
- Allow only authorized people in the construction area.

Some Permits
Weld/Burn /Open Flame
Hot Work (Electrical)
Concrete Coring
Excavation
Confined Space Access
Operation Work Entry
Equipment Movement

H. Contractor Progress Reports

Depending on the work to be performed and the length of the contracted service, contractors may be required to submit reports as evidence of the completion of work activities.

- Minutes of meetings (when held)
- Toolbox meeting records (weekly)
- Man-hour and Injury Report (weekly)
- Schedule Updates
- Material Certifications (as needed) e.g., concrete, stainless steel, cable
- Conformance to Code Statement (if needed)
- Vendor Inspection Sheets (if requested)

The Contractor's key safety responsibilities during the work phase:

- Abide by their approved safety program.
- Be reasonably satisfied that when arranging for subcontractors to carry out work, they are competent and recognize the safety provisions they are expected to follow.
- Ensure required safety training is completed.
- Allow only authorized people in the construction area.
- Display required safety postings.
- Identify the hazards of the work, assess the risks arising from these hazards, and define methods to control them.
- Report accidents and injuries.

I. Inspections and Audits

Both the contractor and the host employer should conduct regular inspections of the workplace to ensure that housekeeping is being kept up and that no new hazards are being created by the work in progress.

J. Beneficial Occupancy

In some facilities a process is in place where space cannot be reoccupied following contracted work prior to the approval of the Fire Department or Fire Safety Engineering. Approval to reoccupy space is granted contingent upon the verification that life safety compliance requirements have been met and that emergency systems have been tested for serviceability and are operating properly.

K. Final Acceptance Determination

Once advised by the contractor that the work is complete, it is in the interest of the field engineer to determine that the originator is satisfied with the end product. The field engineer will also wish to ensure that the following (where applicable) are completed as well.

End of Work/Project Documentation

- Punch-list closed
- As-built/Shop drawings submitted
- Specification changes
- Acceptance Criteria Listing completed (if required)
- Equipment Acceptance Tests completed
- Maintenance training & signoff sheets submitted
- Maintenance manuals delivered.

L. Contract Closeout

Upon satisfactory completion of the work/project the field engineer should inspect and verify that the contractor has left the job site in a safe condition and that all safety-related features are operational in accordance with approved drawings and specifications.

The field engineer may wish to call in the facility safety representative to verify that there were no outstanding safety-related issues and obtain their signature on an Inspection and Acceptance Memorandum generated by Procurement to release the contractor from the project and pay the retention.

A closeout meeting with the contractor should be conducted at this time to review performance and discuss closeout comments.

V.   Conclusion

When first presented with this outline of the safety considerations that should be included in a project the reaction of many organizations is, "If we have to follow all those steps we will never complete a project." However, when all the project team members, which include the originator, the project managers, field engineers, safety personnel, etc. are assembled and asked to help define the sequence of events necessary to complete a project, they generally develop a similar structure when all is said and done. In fact, it is a good idea to invite all the project team members together before a large undertaking to agree to the sequence of events and the involvement each member would like to have at each step. In this manner there will be a consensus of what must take place and who is responsible for its successful execution.

Chapter 4

Roles and Responsibilities

I.   Introduction

The safety of all employees, engineers, managers, subcontractors, visitors, and bystanders in the vicinity of contracted work should be a significant concern to everyone involved in the construction process. Only through the clear definition of responsibility and accountability for safety can personal injuries and other accident-related losses be minimized in a continually changing environment such as that on a construction site.

Responsibility for construction site safety has been the focus of heated debate for some time. Plaintiff and defense attorneys argue the subject regularly. Architects, engineers, and clients often contend that the development of the project's safety program is the responsibility of the construction manager[1] or general contractor.[2] Construction managers attempt to skirt the issue of

---

[1] Construction Manager – The single entity that will be responsible for providing controls over contracts awarded, seeing that they are within the estimated budget, and administering construction without having to engage any field employees or tradesmen.

[2] General Contractor – Contractor who has subcontractors to do some part (or all) of the work that he has undertaken to do for the owner. Previously the term referred to the contractor who employed workers of different trades, and who undertook to do most (or all) parts of the work as directed from a specialist (trade) contractor, who normally undertakes the work of only one trade.

responsibility for safety by arguing limited contractual authority. General contractors point out that their contracts place responsibility on the subcontractors to follow OSHA regulations and that the workers must follow safe work practices. In turn, subcontractors and craftsmen point back up the line to both the client and the general contractor as being the ones responsible for coordinating the work between contractors and for providing a safe job site. To further complicate the issue of safety responsibility, many contracts contain indemnification clauses requiring the contractor doing the work to defend the party letting the contract against third-party claims.

Responsibility for safety on construction sites tends to be confusing because lines of responsibility are often blurred by the attempts of those involved to transfer responsibility and accountability for safety to others. The issue of who is responsible for safety is becoming more contentious as the cost of accidents, insurance, litigation, workers' compensation, and other associated costs escalate.

II.   How is a Safe Job Environment Established?

A safe work environment does not materialize by itself by default. To establish a safe work site the roles and responsibilities of each project participant must be clearly and unambiguously defined. This is particularly so where one party assumes multiple roles in the construction process, such as in the case of an owner who designs and chooses to oversee the execution of the work by multiple prime contractors. What is the role of the owner? How much of the safety program is the owner responsible for? Who is responsible for the safety inspections? Whose job is it to address identified safety hazards? Frequently, the task of assigning responsibilities for safety is overlooked by the owner in the initial enthusiasm and accelerating momentum of a new project.

An owner has two options with regard to establishing a construction safety program, either develop and manage the safety program directly or have it developed and managed by a second party such as a construction manager (CM). In the first case, the owner must have technical expertise and resources to manage the safety program directly. Typically, owners do not have individuals on staff with the technical expertise to direct or manage construction safety programs in a contracted work environment. Where owners do not have the technical resources in-house they generally opt to retain a CM. The CM acts as the owner's agent and manages the work, including safety. In this scenario the owner must clearly communicate its expectations regarding safety responsibilities and how performance will be measured. The owner might go as far as to define specific program elements that are important to their particular culture or define the technical selection criteria that are to be applied to the selection of contractors.

Regardless of the type of approach chosen, the owner/client must adopt and support certain basic tenets, such as

- support of the safety program, including adequate financing,
- define minimum expectations of the project's safety program,
- recognize that construction safety is part of an ongoing process where decisions must be made in a timely manner,
- address problems as they occur,
- verify, via audit, that the safety program is being effectively implemented.

The establishment of a safe work site requires that those in control commit to the importance of having a well-defined safety program during the project's conceptual development. It requires them to keep and communicate this commitment through contractual negotiations, work implementation, and finally through to project completion, never wavering in their commitment to safe work practices.

III. Safety Management

Everyone likes to play sidewalk superintendent and watch those bright yellow pieces of earth-moving equipment arrive on site and when they initially swing into operation. Monitoring construction trades and contract compliance is not as much fun, but is much more important. Usually, it is only after a tragic, personal injury or following a serious safety-related loss that the host employer begins to take an interest in safety. At this stage it is difficult and often very expensive to implement an effective safety program.

We all like to watch construction equipment begin to prepare a job site.

We often mistakenly assume that individuals who have been assigned supervisory and management positions, and contractors who bid to perform work, are all mystically anointed with safety awareness and insight. This is not so. Safety awareness, the ability to recognize hazards, and the technical background to be able to eliminate or control identified hazards is learned. Further, safety sometimes seems to be practiced only when it is clear that the site host considers safety to be important. Perhaps it boils down to the fact that individuals strive to perform well in those areas where they know they are going to be measured.

The site host must define the expected outcome of the safety program and assign responsibility for managing it even before developing the contract language. Safety is just as important as schedule, quality, material control, or any other facet of construction. Like schedule and quality, safety does not just happen. There must be a clear and unambiguous description of the safety program criteria, as in any other aspect of business. Perhaps most important is the definition of responsibility for safety. As there is not generally a contractual relationship between subcontractors or even between prime contractors, it is the role of the site host to clearly establish responsibility for safety down through the organizational hierarchy and between contractors.

There are a number of classical organizational relationships in construction, depending on the familiarity of the client with construction practices and the complexity of the work to be constructed. There is no correct organization, but for the sake of discussion we will work with the following organizational relationship, Figure 4.1. This structure is typical of when the client retains a CM to act as his agent to manage a project. The CM provides leadership for the project, provides technical guidance, and monitors conformance with defined specifications. The construction manager is responsible for integrating the skills and performance of the participants into a cohesive project.

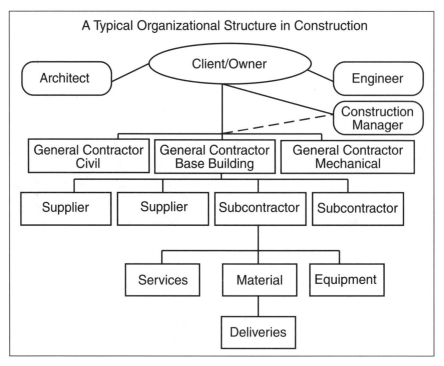

**Figure 4.1 A typical organizational structure
in construction**

IV.   The Client

Clients must clearly define their safety expectations to the project designers and CM, just as required quality standards and the scheduled date of completion of the project are defined. The client must regularly reinforce his commitment to safety by addressing the subject at meetings and when following up on design issues, so everyone involved in the project realizes he is serious about his commitment to safety. The client should require that the designers conduct constructability reviews of their work and that the CM regularly report on the safety performance of the constructors.

The client should periodically conduct its own assessment of the effectiveness of the CM's implementation of safety program requirements and the degree to which safety is being integrated into routine work practices.

V.   Architects and Design Engineers

Included in the architect's and design engineer's (designers) responsibilities should be the assurance that the design they develop can be constructed, operated, and maintained safely. Designers do not generally address construction worker safety for a variety of reasons. They are not typically educated or trained to address worker safety. They claim that they do not have the tools or information to help them design work safety. Probably the real reason is that their legal advisors and insurance companies would prefer that they not address the issue to minimize the designer's liability exposure.

However, design engineers' professional responsibility as outlined in their code of ethics states:

"Engineers, in the fulfillment of their professional duties, shall: Hold paramount the safety, health, and welfare of the public."[3]

Construction workers are members of the general public and are unique facility users clearly at risk when building a structure or system developed by a designer. Designers should show as much regard to the safety of the contractors assembling their work as they show to the facility's end-user.

Architects and design engineers often disavow responsibility for consideration related to safety issues associated with the construction of their work. This will only change if clients insist that designers address construction safety concerns. The following is an excerpt from the deposition of an architect in a construction-related fatality case. Although this is certainly not the position of all architects, it does seem to represent the opinion of some in regard to safety and succinctly highlights the issue of lack of sense of responsibility or accountability for the safety of the construction worker:

---

[3]   The first Fundamental Canon of the National Society of Professional Engineers Code of Ethics (NSPE 1996).

57

[4]**Q.** Are you familiar with the clause in the AIA contract, specifically 13 under general notes, it says, "The general contractors and all subcontractors are responsible for compliance with the provisions of the Occupational Safety and Health Act."

A. Yes.

Q. What do you do, if anything, to ensure the fact that the general contractor and the subcontractors are fulfilling their obligations to comply with the provisions of the Act?

A. Nothing. That was a way of saying that I wasn't responsible for it, that it is a note to protect—in my view, to protect me (referring to a clause the architect inserted into a contract requirement of the builder). In my AIA contract, it says that I am not responsible for making sure that the contractor does those things. In my mind when I wrote that note, it was to say that I was not responsible for that.

Q. As far as your background and experience is concerned, have you ever taken any OSHA-certified courses in construction or construction safety?

A. No.

Q. In your school and in your training, have you ever dealt with OSHA as far as its regulations concerning construction safety?

A. No.

Q. As an architect, have you taken any courses in designing buildings and designing them so that they can be built in a safe manner?

A. No.

Q. Is it fair to say as an architect, your job is to design the building and let the trades build it as they feel is safe and proper?

A. Yes. In our contract, there is a paragraph that clearly says they are responsible for construction techniques. Architects don't get involved in that.

Given the fact that this sort of attitude is prevalent in the industry, clients should require that their designers employ a construction safety professional. This individual should review the project's plans and specifications for potential safety hazards associated with the design and the implied means and methods required of the constructors to complete the work. In this way measures can be taken in a timely manner to eliminate or reduce the potential for accidental losses during the construction work. This approach may also reduce hazards to operating and maintenance personnel occupying the completed facility.

The client's expectation that the architect include consideration for constructability and safety of the completed design should be clearly defined at the onset of the project. Unless this is done, the participants in the design and construction process will assume safety is someone else's responsibility. Costs

---

[4] Shimkus vs. Easy Life Construction Company. Deposition of the project architect, pg. 41, 59.

associated with worker injuries and fatalities are borne ultimately by the client. Insisting that safety be included in design considerations will prevent the occurrence of injuries and ultimately reduce construction costs. This requirement to address safety will likely result in higher designer fees to cover added effort and responsibility. However, for the construction industry, the return on investment of the higher design fees will be a safer workplace and fewer litigation and injury claim costs.

VI.   Project and Field Engineers

Project and field engineers should be given specific direction regarding responsibility for ensuring that due consideration is given to safety and the protection of individuals involved in the construction work. In 1996 the New Jersey Supreme Court rendered the opinion that even when an engineer has no contractual obligation concerning safety, an engineer with actual knowledge of a dangerous condition to which a job site worker is exposed needs to act to prevent an injury.

If the engineer does not bring the hazard to the attention of the exposed party, the engineer may be held liable. A similar judgement was also rendered in the case of Carvlho v. Toll Brothers, et al., 651 A.2d 492 (Super. A.D.N.J. 1995). This case follows a similar Kansas Supreme Court case of Balagna v. Shawnee County, 668 P.2d 157 (Kan. 1983). In the latter, the court overruled the argument that the engineer would have exceeded his contractual authority by notifying a worker of an obvious unsafe condition.

The court acknowledged that the engineer did not have a contractual responsibility for safety. However, the absence of a contractual provision imposing such responsibility does not relieve an individual's need to exercise reasonable care to take some action when circumstances present at the job site demand such intervention to protect a fellow human from harm.

Project engineers must understand what "Stop-Work" means and their authority in regard to safety. Ignoring unsafe behavior condones that behavior. Each time a project engineer walks by an action being performed unsafely, the workers may reason, "I did this before and no one said anything, so it must be okay here." Accepting unsafe practices undermines efforts to convince the employee that this job is different from the last one (where perhaps only schedule was a concern).

A concern that must be addressed with project engineers is the issue of liability to which they might be exposed if they become involved in a safety-related issue or identify a hazard. Often, avoidance is the stance of choice as we have seen above. Such a position is based on the false assumption that any action taken to prevent an injury will be used as an example of "control" over work with regard to safety. This assumption is then extended to the concern that an injured employee may claim that anyone who acted to prevent an injury on one occasion has a duty to do so on others, particularly in the injured party's case.

Preventing an injury is the best way to avoid a lawsuit. Thus the project engineer, who is often the client's representative, must document that a safety concern was identified and formally request in a timely manner through the general contractor that the subcontractor rectify the situation. After all, it is the subcontractor's contractual responsibility to do so.

VII. Construction Manager

The CM retained to act on the owner's behalf to manage the development and construction of a project is also responsible for the coordination of the safe execution of the work. The owner should unambiguously define the construction manager's responsibility and authority for safety. The owner should hold the CM accountable for developing requisite protocols and monitoring the safety program implementation.

In the past, single general contractors generally completed large contracted work projects. A project manager controlled the entire construction process from the development of the work packages through the execution of the work by direct-hire craftsmen. As projects became larger and more complex, project owners opted with increasing frequency to employ CMs to coordinate and oversee the work being performed by several general contractors. The CM is expected to ensure that the project is built to the specifications provided by the engineering design or architectural firm.

Safety is a project management consideration that is sometimes overlooked or is consciously avoided. Many project managers are so busy dealing with coordination issues related to the numerous specialty contractor trades, material suppliers, and providers of rental equipment that they are hard pressed to find time to deal with safety. To complicate this situation, CMs are occasionally brought onto projects after the contracts have been awarded and work has started. The CM may be so busy catching up that he may not have the opportunity to develop a good understanding of the potential hazards inherent in the project before safety problems begin to develop.

Another reason for not addressing safety is the concern about incurring liability and the potential of being named in safety-related litigation. "The CM can either face this challenge, or hide his head in the sand. However, no exculpatory clause will assure immunity from liability. Given this choice, the reasonable approach of many CMs is to recognize and take hold of the risks through a pro-active approach, placing the CM in control of the circumstances creating risks of project injury."[5]

The owner is certainly within his rights to require that the CM explain the manner in which he proposes to implement the safety program. The client should also require that the CM periodically report on the results of the safety program and hold him accountable for the results.

Occasionally, clients make the mistake of being too prescriptive in defining safety program requirements and procedures and then attempt to micro-manage

---

[5] R.D. Connor, *The Agent Construction Manager's Liability for Safety Using a Pro-Active Approach to Manage Liability Exposure,* National Construction Management Conference (1991).

the implementation of the program. The CM assigned responsibility for the site safety program should either be given specific project safety goals or the detailed program and procedures they are to follow, and then be given the authority to run the safety program.

With the responsibility for development and implementation of safety programs comes very real liability considerations. Therefore, the CM must be given the authority to enforce the safety program requirements. Some clients make the mistake of assigning responsibility for safety to the construction manager, but fail to confer the authority to stop work, withhold payment, or use other means of leverage to achieve requisite compliance. Failing to include this authority is like giving the construction manager a gun to point at contractors, but no bullets to back up threats for noncompliance.

VIII.   General Contractor

The General Contractor (GC) is responsible for defining the safety practices of the means and methods to be implemented in the execution of the work for which they are responsible and to ensure that their subcontractors implement those practices.

The GC must ensure that their subcontractors are aware of the site safety requirements and the standards against which their performance will be measured. The GCs should review and agree to the manner in which the subcontractors will perform their work as defined in the Job Safety Analysis (JSA) produced by the subcontractors for each phase of the work they are expected to perform.

Neither the client nor the CM should delegate the decision regarding the standards to be met to the general contractors. Doing so will result in a broad disparity in the implementation of the safety program, eventual discord, and circumvention of safe work practices.

IX.   Subcontractors

Subcontractors are expected to supply the labor and tools to complete the work as scheduled and within defined specifications. They are responsible for ensuring that the individuals they bring to work are technically and physically capable of performing the work assigned to them. They are also responsible for ensuring that the individuals they bring to work have the required equipment and personal protective equipment to perform their work safely.

To be assured that this is the case, subcontractors should be required to produce a JSA for each phase of their work. Where their work creates hazards, such as the release of toxic fumes, excessive noise, radiation, etc., these should be documented to be apparent to the CG reviewing the JSA. The JSA should also define how employees will be protected from the hazards, and the means by which the subcontractor proposes to alert others who might be exposed to those hazards.

X.   Craftsmen

Last in the hierarchical chain are the craftsmen. They are expected to apply themselves and the tools of their trade to produce work of a defined standard. They are expected to perform their work in an informed and safe manner by complying with accepted safe work practices.

Craftsmen are responsible for their own safety in regard to the work they are performing. They must understand that it is their responsibility to ensure that their tools are in safe working condition and that they have the knowledge to perform the work safely. The craftsmen must inspect their own equipment, such as ladders and scaffolds, regularly for obvious defects. They must be aware of site-specific requirements for such things as work entry permits or where the site host insists on placing the first lock on lockout/tagouts. Much of this information will be available through the review of the JSA specific to the work that they are to perform.

XI.   Conclusion

The owner funding the construction work must clearly establish at the onset of a project's development that safety is a serious consideration. The owner must then support this position throughout the project's development and implementation. When an owner assigns responsibility for safety to a construction manager, the construction manager must define the program criteria and see that the resulting process fosters a safe work environment.

Contractors are responsible and should be held accountable for the safety of the work practices they employ and the safety of their respective employees. Contractors should be required to systematically evaluate the hazards associated with their work in order to protect their own employees, and to implement precautionary measures to prevent other individuals from being affected by these hazards. Those hazards that cannot be eliminated or effectively controlled should be brought to the attention of the general contractor to be communicated to other contractor employees.

Contrary to folklore, taking time to consider safety will not delay a job. In fact, extensive field experience in construction shows that for projects in which safety was considered an integral part of business, work was generally completed ahead of schedule. Those work groups on projects that gave due consideration to safety frequently had to wait for groups that did not have high regard for safety. Organizations that fail to implement a good safety program spend substantial time and effort addressing minor, repeatedly occurring inconveniences.

Chapter 5

Plan and Design for Constructability and Safe Operations

"If a builder builds a house for a man and does not make its construction firm and the house which he has built collapses and causes the death of the owner of the house – that builder shall be put to death."

Code of Hammurabi
King of Babylonia
1800 BC

I. Introduction

Although repercussions are no longer this drastic in today's litigious society, it is not enough just to design an appealing structure. With the increasing trend to include owners in liability suits it is in their interest to insist that their designers, who generally focus on the needs of the facility's users, also consider the safety hazards their designs pose to those constructing the work.

Hazards associated with most contracted work such as falls from heights, health hazards associated with removal of asbestos and lead, and work in excavations are generally recognized. However, the risks associated with other work are less appreciated and the opportunities to reduce those risks are often overlooked, such as designing welding locations or steel connections at easily accessible points. Designers may not be able to eliminate all safety and health risks, but they can make significant contributions to workplace safety, if they are motivated to do so. Owners are in a strong negotiating position, as clients, to motivate the designers to make these considerations.

II.   Project Safety Planning

No one will dispute that planning is an important precursor to a successful construction project. Included in project planning considerations should be safety. This should begin early in the project design cycle, in fact before the development of the facility's design begins.

Why develop a Job Specific Safety Plan?
A Plan

- Is a simulation of the program implementation and helps identify flaws in advance.
- Communicates how pieces will fit together.
- Helps people know what they will be expected to do.
- Helps people understand what others propose to do.
- Is necessary to control changes to the plan.
- Provides a basis against which to evaluate performance.
- Helps judge when performance meets requirements.

The first step to control or eliminate hazards associated with the job site and construction-related work is to identify them. This is done through a *Baseline Safety Assessment.* A baseline safety assessment is a comprehensive evaluation of potential safety concerns that are present or could occur on a project, all the way from the start of contracted work activities through to commissioning the completed work.

To be most useful, this assessment and subsequent analysis should be conducted early in the project prior to the start of the detailed design. The results of this assessment can be used by both the host employer and the designers. Through the design process, judicious site layout and careful planning, the majority of work hazards can be mitigated. Finally, the assessment results will highlight areas where increased coordination of contractor activities may be required.

A detailed evaluation of the conceptual design should address constructability[1] issues present in each work phase. Then the job site itself should be assessed for hazards and potential barriers to the successful completion of the work. Once these have been identified, a strategy to mitigate or control them must be developed. This

---

[1]   **Constructability** – The ability to construct the work safely.

process is the *Project Safety Analysis*. The most effective means of completing a Project Safety Analysis is with the support of an analysis and planning team.

A. The Analysis and Planning Team

The coordinating engineer, the designers, and the field engineers who will be involved in the construction process should be included, where possible, in the baseline safety assessment and subsequent project safety analysis. The team should also include safety specialists, maintenance personnel, and even individuals who will use the completed work. The greater the breadth of experience and technical background of the individuals invited to participate, the greater the likelihood of identifying safety-related concerns and hazards. Their involvement will also contribute to identifying effective controls of the identified hazards.

The hazard identification process begins with the team developing an understanding of the project through a review of the scope of the work and conceptual design. With this background information, a visit to the proposed work site is in order. There is no substitute for seeing the proposed job site first hand.

1. *Site Survey & Analysis*

Whether the anticipated work is a grassroots refinery or the installation of a fume hood in an existing facility, a *Baseline Safety Assessment* must be completed. The assessment should begin with a broad view of the site. Look at the site for general issues as if from a distant perspective, such as from a hot air balloon. The following are the sorts of considerations that should be included at this preliminary evaluation stage.

- Nature of the Work Site
  - Property/site boundaries
  - Work site/equipment configuration
  - Soil conditions
  - Waterways and flood potential
  - Apparent buried utilities and/or services
- The Local Environment
  - Weather and prevailing wind
  - Access roads and streets
  - Surrounding neighbors
  - Above ground utilities, power and gas lines
- Political and Social Considerations
  - Building restrictions/requirements for local design approval
  - Local codes and regulations
  - Project impact on activities in the immediate vicinity
  - Impact of noise and increased traffic due to contractor activities

Projects with a discrete scope such as modifications to existing operating facilities often include more complex issues than those on grassroots projects. The objective of a baseline safety assessment in this environment is also to identify hazards and possible safety issues that may encumber the successful completion of the work.

In an operating environment, the assessment team should approach employees who regularly work in the area to obtain their perspective regarding potential safety hazards with which the contractors would have to deal. They should also be asked about concerns regarding the potential impact of construction activities in their own work areas and what the impact of known issues such as dust, congested aisle ways, and increased traffic in the dock area may have on their ongoing operations. Have they got any suggestions on how to minimize these issues? Occupants are often able to point out issues that might not be readily apparent to an occasional visitor. Some issues may be intermittent, such as slippery floors due to a leaky roof that would only be apparent if it were raining at the time of the site visit. These unforeseen conditions could have a negative impact on the work. Evaluation of safety inspection reports of that area may also provide insight into other potential site-specific hazards. What are the work entry and safety permitting requirements? Existing facilities often have specific procedural requirements that need to be communicated to the prospective bidders to be included in their time considerations and estimate.

2. Phase Planning

Next, consider the sequence of work within each work phase of the project and their associated safety issues:

- Contractor Mobilization
- Site Preparation
- Excavation
- Concrete Form Work and Placement
- Mechanical Installation
- Electrical Systems
- Interiors
- Exteriors
- Roofs
- Completed Facility Maintenance

The review of the work to be done in each project phase should attempt to identify the hazards associated with the most probable "means and methods" that will be adopted by the contractor.

There are those who would argue that the means and methods applied by the contractor are the responsibility of the contractor. This analysis in no way absolves the contractor of his responsibility to protect his employees. The objective of project safety planning is to identify the potential safety issues that will be encountered during the work project so that an effort can then be made to eliminate them during the design phase. In this way those hazards will never be encountered in the first place. Evaluation of the work should also include consideration of the potential effect of hazards generated by adjacent contractors and their effect on one another. For example, what effect will a pile-driving operation have on a trench being excavated nearby? The analysis process also identifies potential scheduling considerations.

3.   Enumerating the Hazards and Developing Controls

As each work phase is evaluated, the potential hazards identified should be collated in some format that is easy to read, such as that in Figure 5.1.

Scheduled Activity ID #	Project Phase 1. Site Preparation	Possible Hazard	Party Responsible
Schedule Activity Numbers	**Site Clearing –** Removal of trees and brush	Potential exposure to vermin and broken glass	Site clearing contractor
	Demolition – Removal of existing condemned building	Asbestos and lead contamination	Demolition contractor
	Site Layout – Power line located over project entrance	Crane contact with power line when entering the project gate	CM to coordinate line burial

Figure 5.1 Project safety analysis - Step 1

Contact with overhead power lines is a major cause of fatalities in the construction industry. The identification of this "obvious hazard" is often assumed to be the responsibility of the subcontractor. Clearly, anyone operating high reaching, material handling equipment near power lines must use extreme caution. However, the fact is that as many as 100 workers are killed each year by inadvertent power line contacts. This would suggest that this hazard is not so obvious after all. Crane contact with power lines is the most common cause of fatalities associated with this hazard, but concrete pump booms, backhoe arms, raised dump truck boxes, and other types of equipment also produce fatal contacts. Clearly, the contractor should plan his work to avoid power lines or make sure that the equipment is configured, when being moved, to avoid contact with power lines.

Although a contractor likely to come into contact with this hazard must plan to deal with it, here is a hazard that can be eliminated by the host employer. The first opportunity to eliminate this sort of hazard is during site preparation and layout. Power lines that are located across frequently trafficked routes can be rerouted, buried, or guarded. Once the hazard is eliminated so is the potential for an associated accident or even a fatality.

Those hazards that cannot be designed out of the project should be noted on the project schedule as to when they are likely to be encountered. As activities that expose workers to recognized hazards approach, the contractor can be reminded of the issues. The contractor's preparatory planning can be monitored to assure appropriate safety measures are being taken.

Power Line Protection

OSHA's Safety and Health Regulations for Construction, Subpart N-Cranes, Derricks, Hoists, Elevators and Conveyors, (1926.550(a)(15)) contains specific minimum requirements for the use of equipment near power lines. For lines 50 kilovolts (kV) or less, the operator must keep all parts of the crane or other high reaching equipment at least 10 feet away from all power lines. If the lines are greater that 50 kV, then the line clearance distance is increased by twice the length of the insulator or the formula listed below.

Line Clearance Distance = 10 feet + (0.4 inches) (# of kV over 50 kV)

A good rule of thumb for deciding the line clearance distance is: If the overhead power line is 50 kV or less, then stay at least 10 feet away. For everything else, keep at least 35 feet away.

B. Order of Precedence

Once a list of hazards and safety issues has been developed, the next step is to eliminate or control them. Considerations of how to deal with identified hazards follows.

1. Eliminate the Hazard Through Design

The foremost alternative in dealing with potential hazards is to eliminate them by engineering them out of the design. For example, assembling the majority of a structure on the ground and raising it into place will eliminate fall hazards. The next option is to modify work practices, so instead of working from ladders the use of mechanical lift devices is prescribed. This will significantly reduce the potential for falls. Chemicals also constitute a hazard. One of the biggest problems in the workplace is educating employees to recognize hazardous chemicals and how to safely handle them. Even common materials such as solvents, strippers, degreasers, lubricants, detergents, disinfectants, pesticides, and fuels can be hazardous. Some employees may not be aware of the hazards because the dangerous materials are components of common trade-named products. Substituting known hazards with less toxic material is a means of eliminating this hazard.

2. *Guard the Hazard*

Hazards that cannot be eliminated through design changes should be isolated by means of guards or barricades. The hazard should be made inaccessible and where this is not possible, nearby personnel should be protected from it. For example, equipment guards provide physical protection from moving parts and dangerous conditions such as pilot lights or energized terminals. Examples of guards include

   a.   Rollover protection for operators of equipment which might overturn.
   b.   Fences and guardrails around excavations in the vicinity of walk ways.
   c.   Interlocks that deactivate operating systems when guards or covers are removed.

3. *Give Warning*

When a hazard cannot be eliminated or isolated, a visual or audible warning should be installed to give people timely notice that a hazard is present; for example, backup and travel alarms, flashing lights, or horns. Audible alarms such as backup and travel alarms are more effective than horns activated by equipment operators. Other alarms include

   a.   Proximity alarms such as those that warn of hazardous conditions such as high voltage power lines.
   b.   Threshold detectors such as those used on large paper cutting machines.
   c.   Motion sensors that trigger a pre-recorded verbal danger warning.
   d.   Signs and labels are also warning devices. They should clearly identify the hazard, what harm will result, and how to avoid the hazard. These signs typically include the word "**DANGER**".

4. *Special Procedures and Training*

When the hazards cannot be controlled by the first three methods, special procedures and training must be established. These should ensure that individuals working in an environment where hazards are known to exist are aware of the hazards, that they understand the requisite work practices to deal with the hazards and that they apply those safe work practices to avoid the hazard. This would be the case where working with electrically hot components is required. However, training alone cannot be considered the sole remedy for hazard control where life-threatening hazards are present.

5. *Personal Protective Equipment*

The last resort is to ensure that proper personal protective equipment is provided and used. Once again, employees must be properly trained in the equipment use. Hard hats, safety glasses, goggles, life jackets, safety footwear, respirators, gloves, aprons, and other protective equipment provide a physical barrier between the individual and potential hazards. These barriers only work if used properly. They will only be used properly if the individuals have been trained in their use.

Armed with this information the project safety analysis can continue to evolve (see Figure 5.2). Information related to the control of hazards can now be developed, warning flags added to the project schedule, and a determination made of the safety information that needs to be included to the contract general conditions. For example, if a building that is to be demolished is known to contain asbestos, a flag would be inserted into the schedule to give the field engineer several weeks' notice to verify that the contractor's workers have the requisite training, respirator fit tests, etc. to deal with the asbestos. Once again, the means and methods applied by the contractor are his responsibility, but in the interest of work expediency it certainly does no harm to remind the contractor of the anticipated hazards associated with work they are about to perform.

Schedule Activity ID #	Project Phase 1. Site Prep	Hazard	Controls	Special Notes	Party Responsible
Schedule Activity Number	**Site Clearing** Removal of trees and brush	Potential exposure to vermin and lyme disease	Fumigate Personal protective equipment	Fumigate only when wind is from east. Advise demolition contractor	Site clearing contractor
	Demolition Removal of existing condemned building	Asbestos and lead contamina-tion	Respirator protection	Verify program documenta-tion	Demolition contractor
	Site Layout Power line located over project entrance.	Crane contact with power line when entering the project gate	Eliminate high risk hazard	Ensure buried lines can with-stand heavy equipment loads	CM to coordinate line burial

Figure 5.2 Project safety analysis – Step 2

III. Project Design Considerations

Designers should incorporate baseline safety assessment results into their design considerations. The designer's next safety consideration should be to eliminate the hazards inherent in their own design. Once the design has been developed, the designers should evaluate the probable assembly and construction sequence of the work to determine if their design can be constructed safely.

For example, designers might develop a design that would lend to the roof metal plate connected wood trusses being assembled on the ground and lifted into place to avoid fall hazards. Alternatively, the designer should bring to the attention of the constructor the Truss Plate Institute recommendations for handling and installing trusses to ensure they are aware of specific safe-assembly specifications.

When the safest option is not reasonable or the hazard cannot be avoided, the designer should then select an option that entails the fewest foreseeable risks. The measures available to the contractors to protect their workers during construction, such as personal protective equipment or temporary edge protection, should not be an excuse for the designers not to eliminate hazards identified at the design stage. In the event the designers are not able to engineer out a hazard they should at least identify ways of reducing and controlling the risks that remain to the host employer, if not the contractor.

The kinds of design safety and construction considerations that are being implemented with increasing frequency are

- Early installation of stairs and handrails that are part of the completed facility design, to provide safe access and fall protection for the constructors.

- Prefabrication of roof structures on the ground and raising them into place to reduce the risk of serious falls.

- Heightening building parapet walls to provide fall protection for individuals who have to access the roof to perform maintenance functions.

- Specifying holes in vertical structural steel members for perimeter cables for fall protection.

- Sequencing the installation of exterior wall coverings such as precast panels on multistory buildings as soon as the framework has been installed and before the other trades begin work.

Stairs and handrails installed early in the steel erection process.

- Substituting hazardous material with less hazardous material.

Considerations that have an impact on the individuals who will maintain the completed facility are also important.

- Rotating valve stems so that they do not protrude into walkways.

- Allowing ample space between operating components so replacement and maintenance of equipment will be less likely to result in personal injuries.

- Recommend non-toxic chemicals of comparable performance. If no known alternatives are available, potentially toxic chemicals in diluted form can be recommended.

Position equipment so it does not create a potential hazard for personnel.

- Providing fixed rails on maintenance walkways rather than requiring maintenance personnel to rely on safety harnesses.

- Including roof tie-off points for the protection of window washers.

Work that is recognized to be hazardous does not generally result in deaths, injuries, or health problems.

It is often the case that design decisions result in a hazard that has to be dealt with by the contractor. Designers may contend that other professionals have more experience regarding construction safety and health issues and that contractors are in a better position to ensure that safe work procedures are adopted. However, constructors often do not have the opportunity to eliminate the hazards related to the construction of the work or the job site.

The designers should highlight unusual designs that may require particular attention by the contractor when considering the detailed method of construction. They should also clearly identify the assumptions and principles of the design that may affect its construction or assembly. Designers must provide information to the constructors on those aspects of their design that the constructor might not reasonably be expected to know.

IV.   Constructability Reviews

Can the facility be built safely? Did the designer of the architecturally spectacular building that features details such as steel columns 32 feet across to be filled with concrete give consideration to the safety of the workers expected to erect the columns or pour the concrete?

As the project's design evolves it should be periodically evaluated for safety issues. Certainly when the design has reached 90% completion, a formal design safety review should be conducted. Here, the planning team, the potential occupants, and maintenance personnel should be assembled to review the design. Can the facility be built and subsequently maintained safely?

Designers occasionally fail to include practical necessities related to facility maintenance in their designs. How will the window washers secure their equipment on the roof of the building? Does the mechanical system's piping provide adequate headroom for maintenance personnel moving in the area? Is there more than one exit for emergency egress? How about the building with the wonderful four-story, atrium type foyer where the windows and lighting are out of reach? How are the light bulbs in the ceiling to be changed where the floor loading capacity has not been designed to support mechanical lifts large enough to reach the ceiling? The owner of that facility must currently pay to have four stories of scaffolding erected each time the light bulbs must be changed or the inside windows washed. Often these practical issues must be brought to the designer's attention. The 90% design review is a good opportunity to point out these issues so that the design can be modified to accommodate the needs of the constructors and maintenance personnel.

V.   Contractor Work Evaluation

At this point in the project's evolution we should have a design that will provide the facility operator with a workplace free of recognized hazards. Job-site hazards and hazards associated with construction that remain should have been included in the information forwarded to the prospective bidders.

The successful bidder should be required, by contract, to provide the host employer with a list of selected controls for each identified hazard as well as those hazards identified by the contractor as associated with his specific work. This information should be discussed with the contractor and the controls agreed to prior to the contractor being issued a release to begin construction work.

VI.   Summary

Have you ever driven by a building that you finished years ago and your first thoughts are "That's the building where John died" or "That's the project where Jose became a paraplegic." Often job site managers don't consider planning for safety until the project is well underway or some major calamity occurs. By this time many opportunities to control potential safety hazards are lost or are much more costly to address. Including safety in the initial design is cheaper than negotiating changes with contractors, trying to implement safety controls on a project once it is under way, or after the work is complete.

Chapter 6

Insurance — Demystified

Construction insurance is the third highest, single construction project-related cost, following materials and labor. A study conducted by Stanford University on behalf of the Business Roundtable revealed that insurance premiums on typical industrial projects cost 7% of direct labor costs for workers' compensation insurance and another 1% for builders' risk and liability insurance. Since labor usually represents about 30% of the total project cost, accident insurance costs represent 2.4% of the total project cost.[1] Owners or contractors who are able to transfer responsibility for loss exposures to another AND have that other entity pay for the insurance to cover that responsibility are able to substantially reduce their own loss-related costs.

While some risk managers would advocate the transfer of risk and support insurance as the sole solution to loss control, this does not of itself eliminate hazards or moderate unsafe work practices that are the source of the majority of losses. Nor does the transfer of risk eliminate the potential of being involved in litigation or having to defend a third-party lawsuit. The selection of the type of insurance coverage and determination of limits should be a risk-based decision to protect the project and project-related assets from losses which could disrupt

---

[1] *Owner's Guidelines for Selecting Safe Contractors* by Nancy M. Samelson and Raymond E. Levitt, December 1982.

the project schedule and its timely completion or more fundamentally, the operational viability of the enterprise. However, once again, insurance should not be considered an alternative to a safety program.

I.   Introduction

The subject of insurance is addressed in books concerning just insurance. The purpose of this chapter is limited to providing insight into some of the types of coverage available to the construction industry and is certainly not intended to advocate one type of insurance over another. To begin our discussion of insurance, we should recognize that there are three distinct categories of accidental losses from a risk management perspective:

First, there are the numerous small losses that all projects experience annually whose aggregate costs have minimal financial impact. From a safety perspective these are first aid and minor medical treatment cases. It is generally more efficient for the project to absorb the costs related to these occurrences as part of the cost of doing business rather than recover them through standard insurance claim processes.

The second group of accidental losses involves less frequent occurrences resulting in larger individual losses, with larger annual aggregate costs. These losses may take the form of incapacitating personal injuries, vandalism, or major equipment damage. The most common means of recovery from these sorts of losses is insurance.

The third category is catastrophic loss. These instances happen infrequently; however, their cost is so great that their occurrence could affect the liquidity and even solvency of the business enterprise. A major refinery fire resulting from a contractor error or severe damage of an almost completed facility resulting from a hurricane are examples of catastrophic losses. Insurance is the preferred solution to these occurrences as well.

II.   What Is Insurance?

Insurance is simply an agreement that for some financial consideration another organization, typically an insurance company, will cover potential, predefined losses. Insurance companies, not being philanthropic organizations, will attempt to quantify your potential loss claims based on your firm's historical losses, the nature of your operation, and the effectiveness of your safety program. They will then estimate the administrative cost of handling those loss claims and build in a profit. This will be the basis for their determination of the cost of providing you with insurance and becomes the basis for the insurance premiums you will be charged. The total of the funds collected in this way from all its clients by the insurance company provides the pool of funds to cover claimed losses.

Cost of Insurance =

Expected Losses based on Average Past Loss Experience
+
Administrative Costs
+
Profit

Relying on insurance as the sole remedy to cover losses resulting from construction accidents is a very costly loss-control approach. It offers a poor return on investment as compared to the decrease in loss exposure achieved by directing that same money into eliminating or controlling hazards. As seen in the formula above, in the long run the insurance company recovers the entire cost of any incurred loss, it earns an administrative fee for handling the claim and builds in a profit to boot.

Eliminating or controlling hazards reduces the potential for losses and therefore, the need for insurance. As long as hazards are present, the potential for a loss is present. When an accident does happen, more often than not, one of the accident investigation's recommendations will be to eliminate or more effectively control the hazards that caused the occurrence. Should there be an injury resulting from a known hazard, all the parties aware of its existence run the risk of being named in an associated lawsuit and certainly, the hazard causing the injury will then be mitigated. The logical conclusion here is that the project safety analysis and subsequent safety plan should endeavor to identify and eliminate high-risk hazards.

A. Required Policies

Very little insurance coverage is required by law. The only insurance coverage typically required is workers' compensation and vehicular liability. Workers' compensation covers medical costs of employees injured while on the job, disability and for lost wages. However, there are many other types of insurance coverage to be considered that can protect an organization from irrecoverable losses. A qualified insurance broker can explain the specific details of coverage provided by each type of policy and their related costs.

III.   Comprehensive General Liability

General liability exposures vary widely. In construction, general liability exposure arises from the fact that a work site exists, construction operations are taking place, and independent contractors are present and performing work. Insurance for this type of operation is termed Comprehensive General Liability coverage.

A. Operations & Premises Liability

This specifically defined insurance coverage, under Comprehensive General Liability coverage, is for legal liability for damages resulting from bodily injury or property damage caused by defined occurrences, subject to stated exclusions. This includes the premises where construction work is being performed or operations are being conducted by the insured party. The exposures here are bodily injury or damage to someone else's property.

Premises Liability

A strip mall under construction offered an attractive playground for children of neighboring homes. A young child fell into an unprotected open foundation sustaining serious injuries. Premise coverage provided for the defense of the contractor and settlement of the ensuing bodily injury claim.

Premises Liability

The attorney representing the interest of the injured child visited the project office to discuss the liability claim. When leaving the project's office trailer, the handrail on the steps to the trailer gave way resulting in injuries to the attorney. The contractor's Premise insurance also paid the attorney's claims.

Operations Liability

Operations coverage similarly defended and paid for bodily injury to a residential inspector caused when a contractor's employee struck the inspector when he dropped a load of debris off a roof.

Operations Liability

An electrical contractor improperly connected some wiring in one mall unit. The overloaded circuit resulted in charring the walls and damaging the occupant's electrical hardware. Except for the expense of redoing the wiring, the contractor's Operations coverage paid for the damaged walls, the occupant's electrical hardware, the loss of the use of the premises, and consequential damages resulting from business interruption while the repairs were made.

Operations Liability

New employees unfamiliar with the proper operation of a rented crane caused it to strike the wall of an adjacent structure, dropping and damaging a bundle of siding, and damaging the crane.

No coverage is provided here for the siding, which is clearly in the contractor's care, custody, or control, nor for the rented crane, but damage to the adjacent wall is covered under Operations Liability. An Installation Floater policy would take care of the siding, subject to any applicable deductible. A Contractor's Equipment Floater policy would protect the rented crane.

IV.   Independent Contractor's Protective Liability

This covers the insured party's legal liability for bodily injury and property damage caused by an occurrence resulting from operations performed for him by an independent contractor. It also includes occurrences associated with the insured party's general supervision of that work. When a Comprehensive General Liability policy is issued to a contractor, automatic insurance is provided for the liability that may result from subcontracted operations. Owners may opt to require that general contractors name them, their designers, and their general contractor as Additional Insured under the contractor's Comprehensive General Liability policy instead of writing a separate policy. This requirement should be clearly stated in the owner's contract documents and the owner should request evidence of such coverage being in place before work is authorized to begin.

This insurance is important to cover defense-related costs in today's construction world where it is common practice to sue anyone even remotely associated with the work. Further, in an environment where judgments may be awarded against the owner, general contractor, or construction manager even though the work was sublet, this coverage becomes advisable.

Protective Liability

When a piping subcontractor dropped a wrench on a building inspector, a lawsuit was filed naming the owner, the general contractor, and the subcontractor in the complaint. The owner was defended by Protective Liability coverage under the Additional Insured endorsement. The general contractor's own Comprehensive General Liability policy defended him against liability arising out of the operations he subcontracted and the piping subcontractor's basic Operations Premises insurance took care of his defense, ultimately paying the loss.

A. Completed Operations Liability

A rider to a Comprehensive General Liability policy is Completed Operations and Products Liability. This coverage insures legal liability for bodily injury or property damage caused by an occurrence which takes place in a completed or abandoned operation, or away from premises owned or rented by the named insurer. It also covers goods or manufactured products, sold, handled, or distributed by the named insured or by others trading under his name.

Products Liability

Because of damage caused by leaks, a plumbing contractor was charged with the expense of replacing part of the piping he installed and with replastering the ceiling and walls. Since he was insured for completed Operations, his insurance paid for the replastering and interior damage. The cost of piping replacement was not insured.

B. Indemnity Agreements

This is where one party to a contract says he will defend, pay for, and hold harmless the other party. Indemnity agreements are contract clauses that "pass the buck" and are called a waiver of subrogation. General Contractors indemnify the Owner, Subcontractors indemnify the General Contractor, and Subs to subcontractors indemnify the Subcontractor, and so on down the "food chain."

Everyone from the top of the line on a job site, and down each tier, wants to have the guy below them waive their legal right to seek recovery from a loss. This is the case even when the guy on top messes up, accidentally or even deliberately. It's like a game of musical chairs where there is only one winner. In the game of insurance musical chairs, it is the loser who pays. In some jurisdictions, the indemnifying party is required to cover all losses even if the party being indemnified is negligent. In these jurisdictions you will want to bargain this in or out depending on whether you are the indemnitor or the indemnitee.

Some insurance carriers routinely provide "waivers" in Liability policies. For other carriers the "waiver" must be endorsed to a policy on an individual case basis. Each state has different statutes dealing with them. Additionally, each state's courts will interpret these statutes differently. In several states, the waivers are not legal at all. Although many states have laws against these clauses, one can always find a lawyer who has found an interpretation that allows clients to find some way to protect themselves. So check your contract and know what you are signing. If there is such a clause in your documents, make sure you have a strategy to deal with it.

Case Scenario — The Joint Venture Shuffle

Assume a General Contractor awards a subcontract to a Joint Venture made up of two separate companies to provide cast-in-place concrete, and the subcontract contains an indemnification clause. One of the Joint Venture partners prepares the rebar work and the other the framing, pouring, and finishing. Now, assume that an employee of the Joint Venture contractor installing the rebar is injured and sues the General Contractor.

When the General Contractor seeks to enforce the indemnity provision, and tenders the defense to the rebar entity, to his great surprise he may find that there is no insurance from the Joint Venture . . . the legal entity who contracted to indemnify. The insurance carrier can say that the rebar contractor did not contract to indemnify.

Many owners and general contractors who award subcontracts to Joint Ventures make the mistake of accepting the insurance of each of the Joint Venture partners. The exposure here is that the Joint Venture itself does not have insurance. The Joint Venture is, in the eyes of the courts, the real contracting party. A Joint Venture is a legal entity composed for a specific purpose. The insurance of the separate parties is not related to the operation of the joint operation. In this case, the General Contractor can be stuck defending the lawsuit and paying the judgment.

C. Builder's Risk

Property insurance to protect the project or building during the course of construction is provided by Builder's Risk. If the work is damaged or destroyed during construction and prior to acceptance by the job-site manager, the Builder's Risk Policy will pay to rebuild. Although the "Work" usually belongs to the general contractor until the work has been accepted, the owner generally will require that the general contractor have such a policy in place. Occasionally, the owner assumes that if the contract document requires it, the general contractor is to obtain the policy that will be purchased. The owner should request evidence that the policy has been obtained before work begins and that the owner is named in the policy as the beneficiary.

Builder's Risk policies have deductibles just as in automobile insurance. Builder's Risk deductibles range anywhere from a couple of thousand dollars to several million dollars on larger projects. In the area of Builder's Risk, it is not a bad idea for the general contractor to pass the deductible down to its subcontractors. If the subcontractor knows he is liable for the deductible, he may pay more attention to loss prevention.

If the general contractor intends to pass the policy deductible down to those who incur the loss, the subcontractors, he should make sure the subcontractor is aware of this fact which may not be explicitly mentioned in the subcontract with the general contractor, due to one of those typical subcontract provisions that make the owner/general contractor agreement part of the agreement. Otherwise the subcontractor may find himself paying a deductible with which he was not prepared to deal.

D. Off-Site and Transit Limits

This is another aspect of the Builder's Risk policy. Off-site and transit limits cover construction materials while stored off site or while in transit to the site. Assume the general contractor has a large number of precast concrete components. The precaster, by the nature of the product, produces panels, beams, etc. well ahead of when they are needed. They may well be stored in a nearby vacant lot where this value exceeds the off-site limit.

This material could also be HVAC units for a large structure that are being trucked to your site; the value of these units might exceed the transit limit.

Review the Certificate of Insurance or the Binder on the Builder's Risk. Make sure it is in place before the work begins. Consider the frustration, the cost and delay of not having insurance funds necessary to replace lost units. Determine the policy limits and plan your operations accordingly. Insert a provision in your contract that prohibits exceeding the limits.

V.   Workers' Compensation

There is a growing crisis in workers' compensation insurance as premiums are being driven increasingly higher by rapidly increasing losses. Plagued with runaway medical costs, widespread abuse of benefits, and rapidly escalating insurance premiums, the competitiveness of all business is being threatened.

Increased attention on occupational health issues such as stress, mental illness, and physical impairment due to exposure to asbestos, lead, workplace chemicals and silica is putting added pressure on an already overburdened system. For owners, who ultimately pay the bill, the increased cost of capital construction means fewer facilities and higher fixed costs at a time when world competition demands lower costs.

Even though workers' compensation laws are governed by individual states and vary from state to state, the laws are essentially quite similar even if costs and benefits are not. Each business must meet its workers' compensation liability in some way. State compliance with this requirement falls into four groups.

Open states — where there are no state fund plans and Workers' Compensation insurance is provided by private insurance carriers.

Monopolistic states are those where Workers' Compensation insurance can only be purchased from a state-sponsored insurance fund. There are eight states and territories where this is the case:

Nevada	Ohio	West Virginia	U.S. Virgin Islands
North Dakota	Washington	Wyoming	Puerto Rico

Competitive states are those where you may purchase Workers' Compensation from either a safety fund or from a private carrier. This is the option in 12 states:

Arizona	Idaho	Montana	Oregon
California	Maryland	New York	Pennsylvania
Colorado	Michigan	Oklahoma	Utah

Self-insurance states – All states except the four listed below permit employers to self-insure. The financial requirements are very rigid and include high bonding limits. Workers' Compensation costs generally must exceed $200,000 per year to consider this option.

Nevada	North Dakota	Texas	Wyoming

Fundamentally, Workers' Compensation insurance premiums are determined by using three elements:

Experience Modification Rate or EMR
The Manual Rate
Payroll Units

A. Experience Modification Rate (EMR)

The National Council on Compensation Insurance (NCCI) formulates the experience modification rate. This organization compiles workers' compensation payroll and injury data from approximately 600 insurance carriers. Contractors who have reached a minimum premium threshold size in their particular state are rated and receive a new EMR each year. The calculation of this rate is based on each contractor's payroll and injury data, actuarial factors, weights and ballast. The resulting EMR for each contractor is sent to their insurance carrier by the

NCCI. This information is used by the carrier to calculate the contractor's annual workers' compensation insurance premiums. Therefore, the EMR is an unbiased means of judging the relative effectiveness of a contractor's safety program and the rate with which they experience injuries.

The EMR is a reflection of an employer's safety performance as compared to the average contractor in their specific specialty. An experience modifier of 1.00 is average. This is the multiplier by which an insurance carrier's base rate is multiplied to determine the insurance premiums paid by their client, the contractor. An employer with above average losses will have a modifier higher than 1.00 and conversely, employers with fewer than average losses will have an EMR of less than 1.00. For example, the workers' compensation insurance premiums a contractor with an EMR of .80 will pay will be 20% less than will the average contractor performing similar work. This fact is an important consideration when one reaches the point in the contracting process of judging the ability of contractors to perform work safely and assessing if, in fact, the contractor can complete the work for the price bid.

A contractor who tells you "I have a bad EMR because my work is so dangerous," is either trying to pull a fast one on you or does not understand what an EMR is. He has a high EMR because he has a poor performance history as compared to his peers in his industry. His high EMR is the result of not placing appropriate emphasis on safety.

A contractor's 1999 EMR reflects his loss experience for 1996, 1997, and 1998. When determining an employer's new rating, the oldest year of experience is dropped and the most recent year added. The EMR formula is designed to account for statistical variations in injury severity and frequency and number of employees. An employer with a large number of employees will usually have accident claims that result in a fairly even distribution of both the number and severity of claims.

In the case of small employers, frequency and severity are adjusted to minimize the unpredictability. With smaller employers, primary weight is given to frequency and secondary emphasis to severity. As the size of the employer increases, statistical variations decrease and the experience modification formula reduces to the simple ratio of actual to expected losses. Applying the EMR to the insurance premium calculation provides clear evidence of the value of good safety performance to the bottom line profitability.

EMRs generally vary between 0.20 and 2.60. It should be obvious who has the competitive advantage. The contractor with an EMR of 2.60 pays 2.6 times more in workers' compensation insurance premiums than does the contractor with an EMR of 1.00. The greater the contractor's EMR, the greater the proportion of the contractor's bid is being allocated to cover insurance costs. Logically a contractor with a higher EMR is being paid proportionally less for material and labor to complete the work than would a contractor with a low EMR. The additional benefit to owners of using contractors with low experience modification rates is the proportional reduction in the number of accidents that will occur on the job site and the lower the probability of being involved in litigation.

There are two areas where the EMR may fail to accurately reflect the contractors' accident experience; in the case of a misclassification and where NCCI has erroneously calculated the rate. There are occasional instances of the insurance company placing the contractor in an incorrect Standard Industrial Classification (SIC) or as an administrative expediency. This occurred on one of my projects. An Illinois forestry contractor had been grouped with carpenters, as he was the only such contractor in the state at the time. At the same time, forestry employees were experiencing relatively high injury rates as compared to carpenters. The company's EMR was, therefore, higher than 1.00 since its employees experienced more injuries than those of an average carpentry contractor. The insurance company explained that they were not administratively able to place the forestry contractor in a category by themselves.

The second case is where the EMR is incorrect, based on erroneous payroll and injury data. The reason for this is that contractors often misunderstand the information they are required to report. For example, in some states claims are reported to NCCI "net" of deductibles paid by the employer. That is, the employer pays a selected amount out of his or her own pocket toward the claim. The employer might choose to pay $500, $1,000, or more toward each injury. Where the injury costs are reduced by this amount in the NCCI calculations, it can result in a lower EMR than for a contractor who chooses not to take advantage of this benefit.

On occasion the errors are the result of poor information supplied by the insurance carrier to the NCCI. The insurance carrier may fail to note the deductible paid by the contractor, or provide an estimated payroll value or an incomplete payroll that misses an entire classification, resulting in lower than actual payroll information. The insurance carrier's claim handling philosophy can also have an effect on the contractor's EMR, based on how injury claim reserves are calculated.

If a contractor does not have someone who understands insurance and works with their carrier to ensure that their EMR is calculated properly, the contractor may be conveying the impression that his firm is not concerned with safety and be disqualified from bidding when he well might be a safe contractor.

B. Workers' Compensation Rates

Workers' compensation rates are determined based on the evaluation of losses that occur in each SIC category. The rates are calculated by determining the total cost of all claims for each specific classification code divided by the total amount of payroll for the classification code and expressed as a $ rate per $100.00 of payroll.

Since manual rates are based on work classification, they directly reflect the risk and exposure involved in each type of work. As you would expect, they vary widely from craft to craft. Insurance for a pipe bender is different from that for an ironworker erecting steel. Manual rates, in addition to variation in class of work, also vary widely from state to state. Generally this results from the state's statutory benefits, the level of benefit abuse, regional medical costs, and level of litigation as well as political factors.

Manual rates for the same trade vary from state to state.

The variation from state to state is dramatic. In 1996 the manual rates across the U.S. varied from a low of $14.00 in Nebraska to a high of $103.85 in Minnesota for structural steel erection. As you might expect, clerical workers are not injured on the job frequently and when they are injured their injuries are generally not severe. Conversely, accident statistics for steel erectors reflect that they experience accidents more frequently and the resulting injuries tend to be more serious. The 1998 rate for clerical workers in Illinois was $0.27 per $100.00 of payroll while the rate for steel erectors in Illinois was $32.62 per $100.00 of payroll.

The manual rate reflects the cost of covering the workers' compensation costs for each trade in each state. It does not reflect a single employer's accident history. The premiums of every employer using similar crafts in the state will be charged the same manual rate for those crafts. The variation between states is based on the benefits and entitlements in the local compensation laws.

Example: *Comparison of Contractors*
<u>Contractor "A"</u> *has an effective safety program and good performance history. He enjoys a favorable EMR of 0.60. Assume his direct payroll is $10 million and he has a composite manual rate of $15.00. His workers' compensation insurance premium, modified for good performance, is $900,000.*

$$\text{WC Premium} = \$10,000,000 \text{ ¥ } 0.60 \text{ ¥ } \frac{\$15 \text{ Manual Rate}}{\$100 \text{ of Payroll}}$$

$$\text{WC Premium} = \$900,000$$

<u>Contractor "B"</u> *has not implemented an effective loss control/ safety program except to stay clear of OSHA. His attitude is "we are in a dangerous business, we should expect some people to get hurt." Because of a continuing stream of costly injuries he has developed an EMR of 1.40. He has the same payroll, in the same line of work as his competitor, Contractor A. His workers' compensation premium, modified for poor performance, is $2,100,000.*

$$\text{WC Premium} = \$10,000,000 \text{ ¥ } 1.40 \text{ ¥ } \frac{\$15 \text{ Manual Rate}}{\$100 \text{ of Payroll}}$$

$$\text{WC Premium} = \$2,100,000$$

The difference in insurance costs between the two contractors is

$1,200,000 = $2,100,000 – $900,000

This is 12% of his direct labor costs! The magnitude of savings, of course, will vary from state to state, but even in states with moderate workers' compensation costs the savings is significant.

VI.   Wrap-Up Insurance Programs

The term "Wrap-Up" is being heard with increasing regularity in construction insurance-related discussions. Most contractors have heard of the concept, which has been available for the past fifty years. However, it has only been during the past thirty years that it has become as generally used as it is today.

The term wrap-up is a generic reference to arrangements where a single entity furnishes insurance for the entire project. "Wrap-Up" insurance, Consolidated Insurance Programs, or Owner Controlled Insurance Programs (OCIP) are programs where the owner, architect/engineer, general contractor or construction manager and perhaps fifteen to thirty prime and subcontractors insure only their portion of the work through as many different insurance sources. In a wrap-up the policyholder pays the cost of the insurance directly rather than having each subcontractor include insurance costs in their bids. The wrap-up, therefore, eliminates the requirement that the owner police the contractors to assure that certificates of insurance have been obtained and are correct. Given the economy of scale the policyholder can provide insurance coverage to the project's subcontractors at a competitive price.

A wrap-up program offers several financial and administrative advantages to the policyholder. The most significant advantage is the broader coverage and higher limits available because of volume buying power. The centralization and unified on-site loss control functions and claims handling promote uniform loss control implementation.

The policy is job specific, and runs for the duration of the construction project. It can cover workers' compensation, general liability, and builder's risk under a master contract. Individual workers' compensation policies are issued to the subcontractors in order to maintain a record of their experience and payrolls. Wrap-up policies are written for the term of the project plus any extended periods, assuring continuity of insurance policy terms, conditions, and exclusions.

Wrap-up programs provide some unique advantages. The owner/general contractor receives peace of mind knowing that there are no uninsured workers on the job and that there is no difference in insurance coverage and limits among the various participants. Budgeting becomes clear-cut as the insurance cost is assessed consistently. Subcontractors are not incorporating it into their bid costs on a subjective basis. The carrier provides loss control and claim services to the insured's specifications, ensuring a uniform result and more effective overall management of workplace issues, including application of managed care approaches and the possible use of alternative dispute resolution (ADR), which results in further savings.

Wrap-up is also effective for industrial companies with plans for major capital improvements or plant expansions. For example, an oil company with a multi-year program of refinery upgrades and retooling should explore a rolling wrap-up. This takes all the work proposed by the company within a calendar year and puts it in a single insurance program which covers all the contractors and subcontractors at all their work sites. As a result, the owner/general contractor knows that a properly protective insurance program is in place.

The key to management of a successful wrap-up insurance program is the control of

- all project insurance,
- subcontractor insurance in all tiers,
- loss prevention programs, and
- claims management.

Under a fragmented insurance program, the owner cannot be assured that completed operations coverage would still be in force; nor will coverage for losses that are caused by the contractor's negligence, but that occur after the work has been completed. The wrap-up provides for this coverage to continue for a specified period (usually 2 or 3 years) after completion of the project.

Advantages of Wrap-Up Insurance:

Economy of scale
Reduced administration costs
Avoid double coverage
Eliminating cross-liability suits
Wider broker selection

A consistent approach to unique coverage issues is another advantage of the wrap-up program. Professional liability issues arising from the design-build process can be addressed as can potential losses due to local laws. The insured has better control of potential third-party liability issues that may come from Workers' Compensation claims. Some exposures that may not be covered under traditional general liability contracts can be included in a wrap-up program, including pollution, asbestos and lead abatement liabilities.

Until recently, a single project with an estimated cost of $100 million was considered the threshold for a wrap-up. This threshold has been reduced as brokers and carriers have formed dedicated wrap-up teams. Wrap-up programs begin to become economically viable once there are one million dollars of premium plus losses per year. Wrap-up programs have been written and managed for projects as low as $50 million with rolling wrap-up programs at $75 million over three years. However, some states such as Minnesota and Michigan have minimum level requirements of $80 million total project cost.

Wrap-ups not only offer the security constructors are looking for but also can make good economic sense. However, it is important to keep in mind that a wrap-up program will only be profitable where the job-site manager controls losses through a good safety program.

VII. Conclusion

Insurance is one of those items that owners typically do not like to buy, but sufficient insurance can be critical to the success of your project. Without proper insurance you could lose your entire investment as well as the time and effort you have put into your project.

A project manager's decision isn't just to insure or self-insure an exposure. He needs to know what the coverage <u>really covers!</u> Many construction bankruptcies are the result of failure by the contractor to understand the coverage. Most of the Errors & Omissions lawsuits between contractors and their brokers involve failure to communicate what policy coverage gaps exist. Policy coverage is limited, both in amount of money paid at time of a loss and in the events that are covered. Read the "what is covered", "Exclusions", and "what we will pay on your behalf" portions of the policies . . . even the "Workers' Compensation" policy. Most contractors are operating "naked" of coverage or are under-insured for what they presume to be routine operations.

The types and amounts of coverage you purchase must be evaluated on a cost-benefit basis like any other commodity that you purchase. Enlist your accountant and insurance agent to help you review the amount of coverage you need. However, the bottom line is that insurance is not a substitute for an effective safety program.

Chapter 7

Safety in Construction Contracts

In a contracted work environment such as construction there is generally distrust between the contracting entities. The Owner feels that the General Contractor is going to cheat him, cut corners, and try in every way to make as large a profit as possible. On the other hand, the General Contractor is concerned that the Owner is going to nit-pick every aspect of the work, often rejecting the work and requiring that it be redone until the General Contractor goes broke. In such an environment the contract is the means whereby all the parties to the work to be performed define what is expected and commit to what will be delivered at what price.

We identify people by what they do. Mothers . . . mother, policemen . . . police, teachers . . . teach. People who build buildings, roads, and bridges and other such structures are called "Contractors". This is an appropriate label, since all work in construction is based on a CONTRACT.[1]

---

[1] Frank Keres, *Safety in Construction Contracts,* National Safety Council Congress and Exposition (1996).

I.   Introduction

In construction, if work is not called for in a contract, it is not done. If an owner wants a contractor to do something not included in the original contract, a change is made to the contract. In other words, owners tell contractors what they want, generally through contract documents. Contractors perform contracted work and subcontractors perform pursuant to their respective agreements.

Within the construction industry everyone is a party to a contract, whether it is an owner employing contractors, or a contractor hiring a subcontractor or providing construction services. It is ironic how little there generally is in contracts regarding safety and how much of it is vague or indefinite. To bind a contractor to a desired level of performance, the requirements must be clearly stated in the contract language and be in accordance with local, state, or federal standards.

It is important to note that there is a power shift in the relationship from owner to contractor once the contract has been signed. Therefore, if you are an owner, make sure you have defined what you want done in the contract. Do not assume the contractor can or is willing to interpret your intentions. Define what you want as explicitly as possible, particularly in the area of safety.

Unfortunately, there are no magic phrases or clauses, nor is there a single key to a perfect contract. In this chapter we will explore some common practices and methods and define guidelines for the development of contracts with good safety criteria. We cannot cover every phrase or provide miracle suggestions, but we can examine and offer examples of language (see Appendix 7.1) that has been used effectively by others to help you "contract" better. Environmental, safety, and health terms for construction and contracted services may consist of two pages or they may be contained in a detailed supplement with 30 to 50 pages of specific requirements defining site rules, contractor's safety program requirements, training requirements, inspections, personal protective equipment requirements, etc.

II.   The Contract

The contents of contracts may change as owners add or modify clauses to close the gaps they may have identified in previous contracts. Contractors must be aware that what was in the last contract may not be contained in the next one. There are a multitude of requirements that can be included and numerous ways to include them in a contract. Yet, there are some basic characteristics that must be present in each contract and there are some common methods of including them.

The most common provision in construction contracts from the perspective of safety is that the writer, the entity in charge,

The first rule of contract methodology is "The Drafter is the Master of the Contract."

generally states that "I am not responsible for Safety, you are." To fortify this the drafter relies on indemnity and insurance requirements through which he endeavors to pass on responsibility for safety and associated losses to sub-tier contract participants. This is the way it is. However, from our perspective, as discussed in previous chapters, to effectively control losses one must take a proactive stance and clearly define expectations in regard to the safe performance of work. Just as we define what we want built, the material we want it built with, the quality standards to which we want it built, and the completion date, we must also define safety requirements.

A. Where to Start

First, we must understand the general structure of contracts. The most widely used contract form in construction is the standard Form of Agreement Between Owner and Contractor, commonly referred to as an AIA. Within that contract are the AIA-201 General Conditions (or hybrids thereof). There are approximately 23 pages in this document, but only one page deals with safety. Since architects developed this document they have included words in the boilerplate that state that they are not accountable for safety in the construction of the work they have designed. Without any additions or changes, all that many AIA Contracts have to say on the issue of safety is found in Article 10 – Protection of Personnel and Property of the AIA 201 General Conditions, see Table 7-1.

B. Priorities of Contract Documents

To determine what is included in a contract with regard to safety, it is important to know how contract documents relate and affect one another in order to know what to do or which of the multitude of contract requirements will prevail.

For every contractor on a project there has to be a contract. The owner has a contract with the construction manager or the general contractor. The construction manager has his contract with the general, or on large projects the generals. Then the general has subcontracts with his subcontractors and the subcontractors with their respective subcontractors and service providers.

As the owner is typically the first drafter, the owner's contract has precedence. However, there are numerous phrases and insertions that can change this. The general contractor will in most cases have within his subcontract Agreement language to the effect that *"The Subcontractor agrees to be bound to the General as the General is bound to the Owner."* This means that whatever is required of the general by the owner, the general will require of the subcontractor. Additionally, there will be language that states *"The terms of the Contract Documents between the Owner and the General Contractor are incorporated herein and the Subcontractor is bound thereby."* These terms may not be explicitly defined in the general contractor's contract document. The subcontractor is often expected to ask to see those documents to learn what those requirements might be.

Of course, the subcontractor puts language in its subcontract that the terms of its agreement with the general contractor (which include the general contractor's contract with owner) are incorporated therein and the sub-sub is bound thereby, and so forth down the organizational chain. This results in a contract maze and because the contracts themselves are made up of numerous parts we don't have a contract; we have THE CONTRACT DOCUMENTS.

The first thing to establish when reviewing a set of contract documents is which of the numerous documents prevails. There will be a clause or section somewhere that will lay out the priority of the contract documents. This will typically be in the Owner/Contractor contract or the Instructions to Bidders. The subcontractor is bound by all of the documents between the general and the owner.

Typically, the priority of contract documents in the Owner/General Contractor format is

1. The actual written contract
2. Addenda/Change Orders to the written contract
3. The General Conditions of the contract
4. The Supplementary General Conditions
5. The Plans and Specifications (although some contracts will say the Plans prevail over the Specifications)
6. Instructions to Bidders.

C. What is the Contractor Really Bound by?

Some subcontracts say, *"The Subcontractor is bound as the General is bound."* That's really not much help or informative for either the general or the subcontractor. Other subcontracts will say that in case of conflict between the subcontract and the other documents *"The most stringent requirements will apply."* This means that the contract provision enforcing the greatest duty or penalty on the subcontractor will prevail. Still other subcontracts will say that in case of conflict, *"The terms of the Subcontract will prevail."* This is better for the subcontractor, but if the subcontract is silent, then you are back to the original priority of documents.

Do you currently read ALL the contract documents? If you do, that's great. If not, you are taking a significant risk if you sign a contract that you have not read and understood.

So, how do you deal with this complex system of documents incorporating other documents and one document changing the terms of another? First, you must get a complete picture of the entire set of Contract Documents, including those documents that are referred and incorporated. Then, you have to chart out what is really said. Nothing is more frustrating than reading one document and planning accordingly only to read later that another document changes what you just read. It is more productive to read the entire set TWICE, the first time to know what it says and the second time to unravel the changes and interrelationships.

The excerpts that follow contain the extent to which safety is addressed in the AIA document.

Table 7.1

AIA 201 GENERAL CONDITIONS

ARTICLE 10 - Protection of Personnel and Property

10.1 SAFETY PRECAUTIONS AND PROGRAMS
  10.1.1   *The Contractor shall be responsible for initiating, maintaining and supervising all safety precautions and programs in connection with the performance of the Contract.*

10.2 SAFETY OF PERSONNEL AND PROPERTY
  10.2.1   *The Contractor shall take reasonable precautions for safety of, and shall provide reasonable protection to prevent damage, injury or loss to:*
     1.   *employees on the work and other persons who may be affected thereby;*
     2.   *the work and materials and equipment to be incorporated therein, whether in storage on or off the site, under care, custody or control of the Contractor or the Contractor's Subcontractors or Sub-subcontractors; and*
     3.   *other property at the site or adjacent thereto, such as trees, shrubs, lawns, walks, pavements, roadways, structures and utilities not designated for removal, relocation or replacement in the course of construction.*
  10.2.2   *The Contractor shall give notices and comply with applicable laws, ordinances, rules, regulations and lawful orders of public authorities bearing on safety of persons or property or their protection from damage, injury or loss.*
  10.2.3   *The Contractor shall erect and maintain, as required by existing conditions and performance of the Contract, reasonable safeguards for safety and protection, including posting danger signs and other warning against hazards, promulgating safety regulations and notifying owners and users of adjacent sites and utilities.*
  10.2.6   *The Contractor shall designate a responsible member of the Contractor's organization at the site whose duty shall be the prevention of accidents. This person shall be the contractor's superintendent unless otherwise designated by the Contractor in writing to the Owner or Architect.*

III. Safety Clauses

*10.2.1 "The Contractor shall take **reasonable** precautions for safety of, and shall provide **reasonable** protection to prevent injury or loss to:"*

What does this paragraph in Article 10 of the AIA 201 contract form mean? The word *reasonable* is subject to interpretation and is not definitive. What the owner thinks is *reasonable* the contractor might not agree. As the owner you may wish to replace the word *reasonable* with the word **all** or the word **necessary**. This puts a more stringent burden on the Contractor. Even the Contractor may wish to remove the word *reasonable*. A judge can interpret the word *reasonable* to fit the situation. Apply reverse legal logic here. If an accident happened, and we know that all accidents are preventable, the logical conclusion would be that *reasonable* precaution was not taken or protection provided.

Other adjectives to look out for are "routine", "normally accepted" or "as common in the industry". These types of words and phrases are indefinite qualifiers. It is common to think nothing of this type of language when reading through a contract, but if you think about it, they leave you exposed to someone else's interpretation of how much effort should have been expended in providing a safe work environment. With this type of language in your contracts, you are not enforcing any objectivity with regard to safety. If you are asked to sign a contract with words like this you will be subject to someone else's interpretation, which brings us to the subject of Objectivity and Specificity.

IV. Objectivity and Specificity

Although many owners and general contractors have detailed written safety programs, few of them impose these standards on their subcontractors directly. Consider including the written safety program as a specific contract document. This specificity will increase your ability to ensure work is conducted in a safe manner. Appendix 7-1, Safety Provisions, provides an example of the specificity with which safe direction is provided on some job sites. Clearly, as demonstrated in this example, the safety requirements of each job site will depend on the work to be performed.

Stating that it is a requirement to comply with OSHA is not specific enough. Clarify those areas where there is potential for argument or vagueness. Insert site-specific safety provisions in the contract language such as 'fall protection is required for all employees working above 6 feet'. If you require that the contractor have a drug-testing program in place, define the requirement in the contract. Insert your safety requirements item by item into all of your agreements as a separate addendum to the Contract.

A. Comply with all Laws

As in Article 10.2.2 there is typically a provision that states "comply with all laws, etc." This is an example of an incorporating provision. The "comply-with-all-laws-clauses" suggest that sub-tier contractors are expected to comply with all local, state, and federal safety rules and regulations. This isn't bad, but is it enough?

Until 1997 what did OSHA really say regarding fall protection for ironworkers? Well, it was rather general. In fact, the interpretation of the requirement was such a problem that to clarify the ambiguities in 29 CFR 1926.750 regarding fall protection requirements for ironworkers, the Steel Erection Negotiated Rule Advisory Committee (SENRAC) was established. After protracted discussions and a considerable period of time this committee, consisting of experts from across the United States, was able to reach an agreement. They have since defined the requirements for fall protection for work with steel erection. To say in a document "to comply with OSHA" is often too open to interpretation. Putting this phrase in your subcontract is not going to ensure that safe work practices are followed or save you from OSHA citations, much less the General Duty Clause, not to mention other potential litigation.

1. Laws of the Jurisdiction

Do you know the laws of the jurisdiction in which you work? For example, in Chicago there is a section of the building code that deals with the structural support of tower cranes. Have an accident with a crane in Chicago and the lawyers will have you in their sights. Try telling a jury that

a. You did not comply with the ordinance because you didn't know it was there.
b. You never read it.
c. The phrase in the General Conditions requiring you to comply with ordinances didn't mean anything and you disregarded it.

One specific statute that appears regularly is the duty to provide lateral and subjacent support to adjoining property. Owners must be alert to their duty to give notice to adjoining landowners and to support their building. If this is the case, make sure you comply or that you have specifically transferred this obligation to the general contractor. If it is your intent to transfer this responsibility to the general contractor, ensure that he is explicitly aware that this is included in your provision that he "comply with all laws".

B. Notice of Injuries

Many host employers include the clause that requires the contractor or subcontractor to send all notices of injuries to the host employer or general contractor. If this is the means by which you expect to learn of occurrences, there is something wrong with your own operation. Your field engineers should be advised of the occurrence of injuries as soon as they happen. But, this requirement is a useful clause to eliminate cracks in communication.

Notices of injury clauses are seldom complied with. So, put some teeth into the requirement. The host employer wants to know of the occurrence of incidents and significant near-hits. Consider inserting the requirement that to get paid the payee must first submit reports of injury, or a statement to the effect that there were no occurrences. Include this in the contract provision that deals with documents necessary for payment, such as lien waivers, etc. In this way an accountant in the office responsible for submitting documents to get money will look up those reports and send them in.

To add even more teeth, consider implementing a $100 or $250 penalty on each report not filed with a pay request. Make it known that the construction site Safety Manager's budget will get credited with the penalty fee money where he points out accidents that subcontractors did not report. The Safety Manager will be motivated to ensure his records are in order and that he is aware of all occurrences and near misses. The penalty will also motivate subcontractors to report injuries. Having notice of all accidents can prevent that sinking feeling and asking yourself "What the heck is this all about!" when you get served with the lawsuit several years later and you have no documentation to defend yourself.

Contractors are required to have workers' compensation coverage and are required to provide accident information to their insurance carrier. Since they must provide a certificate of insurance prior to starting work, you have their insurance carrier's address. Approach their insurance carrier and ask them to give you notice of injury reports to confirm your records are complete and whatever other information they are able to share with you within the confines of dealing with their customer. This will provide one more means by which to capture this information.

C. Hidden Provisions

Let's now address the favorite tool of the devious, the hidden provision. These are typically important provisions placed in a sub-section of the general conditions where they might not be seen because they are one of those documents you never see, but which were "incorporated".

Take for example the host-employer who has a clause entitled "Watchmen" which basically says in five long sentences that the general contractor will provide a watchman for any and all hours that the general contractor does not have staff present on the job site. The last sentence then says "The General Contractor assumes all responsibility for and indemnifies the host employer for all fire, vandalism, theft, or any occurrence caused by a third party or an outside source."

This is certainly not an ethical practice. This observation is included as a note of caution to those individuals who do not always read all the contract language or make the effort to locate all contracts-related documentation. The wording of clauses and what they infer can have costly implications. The General Conditions is an incorporated document which contains a section titled Miscellaneous and Supplementary General Conditions. This section provides the opportunity to change what was said before or state it in a confusing manner. For example, the Supplementary General Condition might say "In Section 10, in line 18 before the word "safety" insert the word "all." This one little word insertion required in a Supplemental General Conditions can change your responsibilities dramatically.

What is Not There

A word of warning, you must do more than read. In EVERY contract you must also determine what isn't there. You might become so caught up in reading and analyzing, that you fall into a VERY COMMON TRAP. The contract documents might NOT include a term or provision that is very important.

Therefore, you should know what you want in your construction documents. If there is no reference to safety or there is no reference to something you assume is normal, then insert it. Make sure what you want there is there.

Take the case of the absence of a clause that the work completion date is contingent on the arrival of critical equipment. The contractor without a clause that defines this contingency may be bound to complete work by a specific date without regard to the actual date of a critical component over which that contractor may not have had any control. Beware of what is not in the contract language.

V.   Summary

Safety should be part of every contract. The most common excuse for not including it is that "Well, if I define safety requirements in a contract, I am accepting responsibility to enforce them; I might get sued."

If you don't have an effective safety program, the probability of being sued is even higher and you will not have an effective defense. Defining safety requirements in your contracts gives you move leverage and gives the contractor no excuse for not having included this consideration in his cost estimates and in his work planning considerations.

APPENDIX 7.1

EXAMPLE TERMS AND CONDITIONS FOR FIXED-PRICE CONSTRUCTION CONTRACTS

ARTICLE 30. SAFETY AND HEALTH

The Contractor shall take all precautions in the performance of the work under this contract to protect the safety and health of the contractor's employees, other workers and members of the public who might be affected by the work. The contractor shall comply with all the applicable safety and health regulations and requirements included in Title 29 of the Code of Federal Regulations including, but not limited to the parts 1910 and 1926, as well as other state, federal and local regulations for construction. The owner shall notify the Contractor, in writing, of any noncompliance with the provisions of this clause and the corrective action to be taken, which may include suspension of employees from the site. After receipt of such notice, the Contractor shall immediately take corrective action. In the event the Contractor fails to comply with regulations and requirements of the clause the owner may without prejudice to any other legal and contractual rights issue an order stopping all or any part of the work. Thereafter, a start order to resume the work will be issued by the owner. The Contractor shall make no claim for an extension of time or for compensation for damages by reason of, or in connection with, such work stoppage. The Contractor shall ensure that all its employees and all subcontractors are aware of and follow the Contractor's approved safety and health program as well as all regulations in this clause.

A. Reporting Requirements

1. All accidents occurring at the facility must be reported immediately by dialing 911 from facility telephones. When using pay or cellular phones, dial (633) 255-1911. Accidents must be reported immediately to the field engineer. In addition, an accident investigation report must be submitted to the Field Engineer within 24 hours. The types of occurrences that must be reported include, but are not limited to fire, explosions, personnel injuries/illnesses, security incidents, vehicle accidents, toxic or flammable material spill or releases.
2. A "Man-hour and Incident Report" which summarizes the number of man-hours worked, injuries, lost-time accidents and number of days lost shall be submitted with the Contractor's request for payment. The report shall be submitted on or before the second work day of the following month, and upon completion of the contract.

B. Contractor Environment Safety and Health (ES&H) Program and Implementation Plan

1. Within ten (10) calendar days after award of the contract, the Contractor shall submit its ES&H program and Implementation Plan encompassing all applicable aspects of Title 29 of the Code of Federal regulations including part 1910, "OSHA Safety and Health Standards for General Industry", and Part 1926, "Safety and Health Regulations for Construction", and as a minimum shall include the provisions set forth in the solicitation documents issued for this procurement. The Contractor is required to comply with the requirements set forth in its plan. The Contractor's ES&H Program and Implementation Plan must be signed by one of their company's officers.

2. The Contractor's ES&H Program and Implementation Plan will be reviewed for compliance with the requirements established above. If found to address the hazards and work known to be associated with the work to be performed the owner will agree to the plan. Otherwise it will be returned to the Contractor with comments on areas which do not appear to adequately address the hazards and work to be performed. Fieldwork will begin only after the ES&H plan is agreed to and a pre-work release is held. Any revisions subsequent to the initial approval shall be submitted and approved prior to the Contractor's implementation of those revisions. The Program and Implementation Plan as a minimum shall include provisions for the following:

   a. Statement of the Contractor's ES&H policy.
   b. The name and qualification of the Contractor's designated ES&H representative and alternate, the names of competent persons for excavation, scaffolding, confined space entry, etc. as required by the work conditions.
   c. The frequency of regular safety inspections to be conducted by the Contractor.
   d. Job-specific orientation for Contractor and subcontractor employees. This shall include
      i. Review of the Contractor's ES&H Program and Implementation Program and job safety analysis,
      ii. Review of hazards and associated precautions identified in the Construction Specification Special Conditions,
      iii. Review of construction hazards which should be emphasized to employees because of past experience or seriousness of the hazard; and approval must be obtained prior to any job activity requiring a competent person,
      iv. Review of emergency numbers, emergency egress routes and assembly points.

e. The schedule of weekly safety meetings to be held with employees to emphasize project safety and health, environmental protection and fire prevention;

f. The location at which the "Occupational Safety and Health Protection Poster" and other occupational safety and health compliant forms will be available to employees;

g. Implementation of specific ES&H requirements listed in the special conditions of Contractor or Specifications;

h. Drug-Free Workplace requirements

i. Disciplinary policy and procedures

3. The Contractor's approved ES&H representative must attend the pre-work release meeting.

4. The Contractor is responsible for reviewing and approving its subcontractors' ES&H program(s) and Implementation Plan(s) that must comply with the requirements of the contract prior to commencement of work on site.

5. If the Contractor has an approved ES&H Program and Implementation Plan on file with the owner, revisions necessary to address new work shall be submitted and approved prior to commencing the new work.

C. Construction Job-Specific Safety Analysis

1. The Contractor must submit within ten (10) days and have reviewed by the designated field engineer, prior to the pre-work release meeting, a job-specific safety analysis which details the specific hazards associated with each phase of the job as well as the mitigating actions the Contractor shall take to reduce the risk of injury. Material Safety Data Sheets (MSDS) are to be submitted as part of this analysis. The analysis as a minimum shall include the provisions set forth in the solicitation documents issued for this procurement.

2. Specific procedures in the areas of fall protection, excavation, confined spaces and hoisting and rigging are required as job conditions dictate. Names and qualifications of competent persons as defined by OSHA and ANSI must be submitted a minimum of seven (7) days prior to the start of those activities.

3. The Contractor shall provide a job-specific safety orientation to all Contractor and subcontractor employees. Each Contractor employee shall sign the job safety analysis to indicate having received the orientation. The signature list shall be submitted to the Field Engineer at the end of the first work day and throughout the duration of the contract when signatures are added.

4. The Contractor will submit a site-specific safety plan reflecting the hazards identified in the Site-Specific Safety Analysis and be integrated with the project schedule to determine the time dependence relationship of the individual hazards.

5. The Contractor will require a safety analysis be prepared by each subcontractor and their lower tier contractors of all work items associated with their portion of the work.

6. The Job Safety Analysis must be reviewed to incorporate any changes determined necessary during construction. The Field Engineer must agree to the revisions prior to the activity taking place. All employees affected by any revisions shall be notified and advised by the Contractor.

D. Safety Documentation

The Contractor shall submit the following documents, certificates, etc. as required.

1. Equipment inspection documentation required by 29 CFR 1926, Subpart N must be with the equipment and shall be provided to the Field Engineer prior to use. This includes personnel lifts, cranes, augers, suspended scaffolds, spreader beams, and lifting devices.

2. If the Contractor intends to administer first aid or Cardio-Pulmonary Resuscitation (CPR), the Contractor must comply with 29 CFR 1926, and include a list of names of employees who will administer first aid of CPR along with a current certification as part of the JSA.

3. The Contractor must maintain Material Safety Data Sheets (MSDS) at the job site. MSDSs for all products and materials brought on site shall be posted at a Contractor's bulletin board that is accessible to all workers on the job site. In addition, all MSDSs must be submitted as part of the JSA.

4. Pressure vessel certificates per 29 CFR 1926.29 must be submitted and approved prior to use.

5. Documentation of employee training and/or proof of proficiency required by OSHA shall be submitted to the Field Engineer prior to the commencement of their work. Examples include CPR certifications, confined space training, competent persons for excavations and scaffolding, and fall protection training.

E. Variances

Requests for exceptions to the Owner's health and safety requirements, Contractor's approved ES&H Program and Implementation Plan, or Contractor's approved JSA must be submitted in writing to the Owner. Exceptions shall not be implemented without prior written approval.

F. ES&H Orientation and Training

All Contractor personnel are to attend an ES&H orientation before starting work at the site. The training consists of two parts: Contractor Safety

Orientation (CSO) provided by the Owner and Job-Specific Orientation conducted by the Contractor.

The CSO orientation lasts approximately one and one-half hours. This orientation is required on an annual basis. Upon completion of the orientation, each employee will receive a wallet card that must be presented to Owner personnel upon request. Failure to show proof of the orientation will result in re-attendance of the CSO the next day that orientation is scheduled.

The contractor must provide a job specific orientation to all Contractor and subcontractor personnel prior to the start of construction activities as described in paragraph B.2.d. above.

The Contractor is responsible for ensuring that their employees and subcontractor employees who do not speak English understand all ES&H requirements. The Contractor must also be able to communicate emergency instructions to those employees.

Contractors shall hold and document the following meeting:

1. Weekly "Toolbox" meeting (5-15 minutes) for all Contractor and subcontractor employees at the site to discuss pertinent safety topics.
2. Meeting minutes or discussion topics must be posted on the Contractor's bulletin board for a period of one month following the meeting. Minutes shall include the date, person holding the meeting, subject covered and signatures of attendees.

G. Equipment and Tool Inspection

All tools and equipment brought on site by the Contractor will be inspected by the Owner for compliance with OSHA and good work practices prior to use. Tools and equipment will also be randomly inspected throughout the duration of the contract. Items found to be out of compliance shall be immediately removed from service, tagged out of service, and taken off site by the Contractor at the end of that work shift.

H. Disciplinary Program

The Contractor is required to develop and implement a disciplinary program to control poor performance, misconduct, negligence and safety violations by both its employees and those of any subcontractors. The program must be reflected in the Contractor's Safety and Health Program. If it is determined that the Contractor has not implemented such a program, the Owner will enforce its own disciplinary action up to and including termination of the contract. The Owner will issue verbal warnings to Contractors and Subcontractors of safety infractions and also will issue written citations for more serious or continual infractions.

1. Imminent Danger

   When an employee is observed to be involved in a situation that places himself/herself or others in imminent danger of being seriously injured or killed, progressive discipline will not be enacted. The employee will be suspended from the Owner's facility for a period of six (6) months. The Contractor's ES&H representative may be suspended from working at the Owner's facility for three (3) days. The Contractor's management will be notified of the suspension by the Owner's procurement office.

2. Contractor's ES&H Representative Status

   The Contractor's ES&H representative may receive citations for failure to enforce safety program requirements. Any Contractor's ES&H representative who received a suspension of any kind will not be allowed to continue in the ES&H representative capacity until reinstated by the Owner. Any suspension evoked by the citation will start on the day following the citation to allow the Contractor time to arrange for a replacement, unless the seriousness of the violation warrants immediate removal from the site. The Contractor is responsible for submitting for approval the name and qualifications of an ES&H representative before work will continue.

3. Notice

   If the Owner disciplinary action results in suspension of Contractor employee(s) or ES&H representative as discussed above, the contractor shall make no claim for extension of time or for compensation for damages by reason of, or in connection with, this disciplinary action.

   A Contractor's safety performance will be an important factor for future consideration for bid lists and selection criteria. This will include a review by the Owner of the Construction Contractor's performance, misconduct, negligence, and safety violations by both its employees and any of its subcontractors. If it is determined by the Owner that the Contractor has failed to implement its approved ES&H program and the contractor has shown negligence in enforcing safety compliance on the Owner's site, the Contractor will be removed from the active bid list of Contractors and shall not be allowed to bid or work as a subcontractor on the Owner's site for a period of time as determined by the Owner.

I. Drug Free Workplace

It is the Owner's policy to maintain a drug free workplace. Possession of illegal drugs, alcoholic beverages, and drug paraphernalia is PROHIBITED. Individuals found with these substances in their possession will immediately be denied access to the work site and the employing contractor's contract could be terminated.

The Contractor and all lower tier subcontractors shall abide by the Drug Free Workplace Act of 1988.

Employees taking medications will notify their immediate supervisor.

Any employee involved in an accident which results in personal injury or property damage may be given a drug/alcohol test. Employees testing positive may be permanently banned from employment at any Client facility. The cost of such testing will be the employer's responsibility.

J. Owner Site-Specific Requirements

The following are requirements that must be implemented on the job site. These must be included in the Contractor's ES&H Program and Implementation Plan.

The Owner conducts work through the use of on-site permits. The Field Engineer shall identify and will arrange for all necessary permits. There is no cost to the Contractor for any Owner permits. No work activity shall be performed without the required permits. Such permits include work entry, working hot electrically, open flame, confined space entry, digging, using powder actuated tools, moving Owner property off site, removing asbestos. The Contractor shall comply with all restrictions or provisions listed on the permits. A permit to bring radioactive sources or x-ray equipment on site must be approved 48 hours in advance.

1. Personal Protective Equipment (PPE)

   Contractors shall insure that their employees are issued and use proper personal protective equipment when needed.

   a. All employees shall wear safety glasses with rigid side shields at all times in construction work areas unless a higher level of eye protection is required for special hazards. All eye protection must meet the requirements of 29 CFR 1926.102. Safety glasses must be ANSI approved and must be marked with the ANSI designation "Z87.1".

   b. Hard hats shall be worn at all times in construction work areas. Hard hats shall meet the ANSI Z89.1 standard as defined by 29 CFR 1926.100 and bear the "Z89.1" designation. High voltage exposure work requires hard hats that meet ANSI Z89.2 standards and bear the "Z89.2" designation.

   c. All employees shall wear clothing suitable for the work and weather conditions. The minimum shall be a short (1/4 length) sleeve shirt, long trousers, and hard sole leather work boots providing ankle protection. In addition, any work that presents a greater hazard to the feet or toes requires the use of steel toes or metatarsal guards. Canvas, tennis, or deck shoes are not permitted within the construction work area.

d. Contractors shall designate and identify a competent member of their organization whose duty shall be the implementation of the Contractor's ES&H program on the job site.

   (1) The Contractor shall submit the names and qualifications of the ES&H representative and alternates to the Owner for approval prior to assignment of duties.

   (2) The ESH representative shall be present at all times work is being performed on site. If the ESH representative must be off site, the Contractor shall notify the Owner of an alternate.

   (3) Duties include, but are not limited to enforcing the company safety program as well as Owner requirements, providing job-specific ES&H orientations, investigating incidents/accidents, making daily inspections and reporting safety related information.

   (4) The ESH representative must have the authority to stop work and change the operation to correct any deficiencies or to eliminate any hazards observed.

   (5) The ESH representative shall have taken, as a minimum, training equivalent to the OSHA 10-hour training course in construction safety before field work at the Owner's site begins. Documented evidence of attendance, signed by the OSHA certified inspector, shall be submitted to the Owner for confirmation.

e. Each Contractor shall designate a competent person to ensure safe work practices are being followed in areas where OSHA requires competent persons. This person must have the ability to recognize safety and health hazards, must have the ability and written authorization from the employer to take necessary action to eliminate unsafe conditions and actions. It is the responsibility of the employer to determine if the individuals meet qualifications based on guidelines defined in OSHA 29 CFR 1926.32 (d) (f) (I) (m). The competent person must be on site at all times during the performance of the work.

2. Safety Inspections

a. The Contractor shall conduct ongoing safety inspections of the work area. A weekly safety inspection report shall be submitted to the Field Engineer.

b. The Owner's Field Engineer, will make periodic safety inspections of construction areas and will document deficiencies when observed. Deficiency Reports will be submitted in writing within 24 hours to the Contractor. The Contractor shall be responsible for replying to the Field Engineer within 24 hours as to what corrective action has been taken on the deficiencies.

c. The Contractor's competent person performing daily inspection required by OSHA, such as trench and excavation, ladder, and scaffold inspections, shall document each inspection. Such documentation shall be signed and include date, time, and conditions found. Documentation shall be available for review by the Owner for the duration of the project.

d. Contractor employees shall immediately correct all unsafe conditions and unsafe acts. Those that cannot be corrected shall be reported to supervision.

e. Unsafe acts by any personnel may be grounds for immediate removal and permanent barring from the job site.

3.   Emergency Response

a. Each Contractor shall provide first aid supplies on site for their employees and a person trained in basic first aid who can render immediate care when needed. The name of the designated first aid provider and a copy of training documentation will be provided to the Owner, with a copy to the Field Engineer. The Contractor shall provide transportation to the medical facility for injured employees with minor injuries. When needed, seriously injured employees will be transported by the Owner's ambulance. The Contractor should not permit an injured employee to drive himself to the medical facility or home, unless approved by a medical professional.

b. All accidents or near-miss incidents shall be reported to the Field Engineer immediately and a thorough investigation conducted and documented.

c. Contractors shall complete accident reports and "Employee's First Report of Injury or Illness Form." A copy of these documents is to be provided to the Field Engineer within 24 hours of the incident.

4.   Tools and Equipment

a. Electric tools and extension cords must be grounded. Ground Fault Circuit Interrupters (GFCI) shall be used whenever electrical tools or extension cords are used. An assured grounding program is not an acceptable alternative.

b. Hand tools shall be maintained in a safe and usable condition.

c. Only authorized, properly trained employees shall operate equipment, machinery, vehicles, and tools. Only the operator shall ride on any equipment unless specifically designed to carry more people.

d. Scaffolds and ladders shall be erected on sound, rigid footing, capable of carrying of carrying the intended load without settlement or displacement. Adherence to the OSHA standard is required. The Owner has a tagging system in place and requires that all scaffolds be inspected and tagged by the Contractor prior to use.

e. Powder actuated tools shall be operated by licensed operators only with the facility Owner's approval. This equipment shall be operated only when using required personal protective gear.

106

f. Metal ladders are prohibited.

g. Use of explosives is prohibited without written approval from the Owner.

h. Personnel lifts must be equipped with audible motion alarms for movement in any direction. All lifts must be equipped with safety foot pedals for operation movement. The equipment manufacturer must approve any modifications to lifting and hoisting equipment.

i. No alarms, safety devices, etc. will be disabled without approval of the Field Engineer. The Contractor shall make a specific request to the Field Engineer 48 hours prior to the shut down.

j. Lockout/Tagout procedures shall be enforced. Owner maintenance personnel will de-energize systems and initiate lockout/tagout. Contractor personnel must be trained in lockout/tagout prior to participating in lockout/tagout of hazardous energy sources and working on lockout/tagout systems or equipment.

5. Fire Protection

a. At least one (1) 10 Lb. ABC rated fire extinguisher shall be within 25 feet of operating power tools (gasoline or electric) any time welding, grinding, or other spark generating work is being done; and any time paint or cleaning solutions are being used.

b. A fire watch shall be utilized any time welding, grinding, or other spark generating work is being done and thirty (30) minutes after work is completed unless a waiver is provided by the Owner's Field Engineer. An open flame permit must be in place prior to any welding/cutting operations and be posted in a conspicuous area at all times.

c. Open burning, fire barrels, or other open-flame heating devices having exposed fuels below the flame are prohibited. Flashback preventers are required on oxygen/fuel hoses.

d. All engines shall be shut off before refueling. Smoking is PROHIBITED around all volatile fuels, vapors, or combustible materials.

e. Plastic containers for storage of flammable or combustible liquids are prohibited. Only approved containers and portable tanks shall be used.

f. Smoking is only permitted in designated smoking areas.

6. Vehicle Safety

a. All Contractor employees who require driving privileges within the project site shall view the Owner's Driver Safety video and Construction Safety Training video. Vehicle operators must have an appropriate, valid driver's license when operating vehicles on site.

b. The driver and front seat passenger of any vehicle shall utilize seat belts any time the vehicle is in motion on Owner property.

c. Operators of equipment shall utilize seat belts at all times when equipment is being operated.

d. All vehicles and mobile powered equipment, except automobiles and pickup trucks, must have backup alarms.

e. If required by the manufacturer, rollover protection structures shall be provided.

f. All vehicles that do not have Owner permits shall be escorted at all times when on an airfield by an Owner-permitted vehicle, whose driver has airfield driving privileges.

g. All motor vehicles operating airside, (except on service roads) shall be equipped with a working MARS light, which shall be operating at all times. The MARS light shall be attached to the top of the vehicle and visible for 360 degrees.

h. Construction equipment operating airside shall have an operating MARS light attached to the top of its cab or shall have an orange and white checkered flag attached to its tallest point. If the checkered flag is used, the minimum size shall be 2' by 2'.

7. Site Security

a. The following are incorporated as part of the Client's Master Security Plan and are adopted as Rules and Regulations which all persons having unescorted access into the Air Operations Area (AOA) at the Owner's facility must adhere to:

(1) Authorized employees must enter the Air Operations Area (AOA) through terminal buildings at authorized portals only. Access is gained by swiping their ID card through a card reader and entering their valid PIN number (personal identification umber) into the pin pad on the card reader. Outer perimeter posts require swiping of ID cards only through authorized portals. Failure to do so may result in denial of future access to the site.

(2) Employees who fail to display their issued ID, or in the case of a flight crew member, their appropriate company ID which allows them unescorted access, on their outermost garment, above the waist, while in the defined secured area or Security identification display area will be subject to Section (f).

(3) Employees who have unescorted access and are within the secured area and do not challenge persons not displaying their issued or approved ID card will be subject Section (f).

(4) Employees who have unescorted access and who enter a secured area access door and allow anyone access through that portal behind them (piggybacking), will be subject Section (f).

(5) Employees authorized to escort personnel in the secured area must have the ability to direct and control the physical movement of the individual(s) being escorted the entire time they are in the secured area. Failure to do so will subject you to Section (1).

(6) Any employees found in violation of Sections (1), (2), (3), (4), and (5) will be issued citations by the Department of Aviation Security Officers that include the following penalties:

1st offense	$50 fine
2nd offense	$100 fine
3rd offense	3-day denial of site access

Fines are to be paid within 72 hours from date and time of issue. Fines are to be paid at the Badging Office during normal business hours, by check or money order. Failure to do so will subject you to revocation of your ID badge.

Failure to comply with these regulations may be considered as a violation of your contract.

Chapter 8

Contractor Selection

Project safety will only be as good as that of the poorest performer on the job site. Selecting contractors with a demonstrated record of safe performance is perhaps the most effective means of improving the odds of having a safe job site. Both project managers and owners are becoming increasingly aware, as they experience the ever-increasing rise in injury claims and accident costs, of the need to be more selective in their choice of contractors. Safety practitioners, owners, and general contractors who regularly use contracted services are advocating the consideration of contractor's past safety performance in their selection criteria.

Rationalizations such as "I am compelled to accept the contractor with the lowest bid" or "I am not permitted to pre-select contractors, so I can't reject a contractor based on their past accident experience," can be overcome and should not be impediments to the selection of safe contractors. Technical selection criteria that include contractors' past safety performance can be used in order to evaluate safety suitability just as quality and technical competency are evaluated.

I.   Introduction

It has generally been the practice in construction to base the selection of contractors solely on the lowest bid. Consideration of safety in the selection process tends to be the exception rather than the rule. Why this continues to be so is difficult to understand.

The Business Round Table, an organization whose membership includes CEOs and COOs of 200 of the leading Fortune 500 businesses, stated in 1989 in their A-3 report[1] that consideration of safety in the bidding process measurably improves safety performance. Further, the report indicated that contractors with good safety performance are more efficient in the execution of their other functions.

Consideration of safety is successfully included in the contractor selection process by enlightened organizations. The question that begs to be answered is why safety is not given more weight in the contractor selection process by more organizations. Surely it makes sense to choose to work with contractors who have a demonstrated track record of good safety performance, thereby decreasing the odds of having accidents on your job site. Additionally, projects that are completed under budget are generally those with lower than average loss rates.

Is it that job-site managers are not aware of the long-term financial impact that safety-related losses have on a project's bottom line? Do job-site managers assume that contractors bear the cost of accidents that occur to their workers on a job site? Don't they realize that contractors offset their accident costs by charging their next client for them? Where else are they going to get the funds to pay for the accidents? Perhaps it is the fact that the financial impact of job-site accidents generally materializes after the field work has ended and the job-site manager has moved to another project so they don't have to deal with the financial impact of accident losses.

The rising incidence of accident litigation and escalating workers' compensation costs is getting corporate attention and they are starting to hold job-site managers accountable. Those host employers who have not yet incorporated safety in their selection criteria are beginning to look for solutions. The following example demonstrates the effect that accidents have on contractors' bottom line.

Example 1

This scenario compares the Net Profit for three carpentry contractors with different accident experience rates. Consider that each contractor is retained for a job requiring 400,000 man-hours of effort ($8 M in labor costs).

---

[1]   Improving Construction Safety Performance, Report A-3, September 1989, The Business Round Table.

Example Company	Lost Time Rate	Lost Time Cases	Average Claims plus Indirect Costs @ $25 K/ Case[a]	Profit Margin 5%	Net Profit
A	3	6	$125,000	$400,000	$275,000
B	5.4	11	$275,000	$400,000	$125,000
C	9	18	$600,000	$400,000	($200,000)

[a] National Safety Council 1997 estimate of Lost Time Case total costs.

The industry average Lost Time Rate for the Carpentry Trade is 5.4 cases per 200,000 man-hours worked. At a 5% profit margin on $8M of labor costs, this would provide each contractor with a potential profit of $400,000. The medical expenses in this example reflect the average cost of lost time accidents on net profits, as reported by the National Safety Council.

Focus on Contractor "C". As this contractor allocates more financial resources to deal with job-related accident costs, fewer funds are available to cover material and labor costs. Not generally being philanthropic organizations, contractors have little choice but to pass on costs resulting from past losses to future customers. Contractors whose workforce experiences high rates of injuries bear the burden of higher insurance premiums than do their safer competitors.

If a contractor has a high overhead resulting from past accidents, where are the savings to be achieved in a competitive bid environment? There are really few areas where savings might be realized:

- cheap (inexperienced) labor,
- distributed (frequently absent) supervision,
- inexpensive (substandard) material, or
- anticipation of numerous change orders and filing of loss claims.

Although none of us want to see anyone lose money, neither do we want to pay for losses contractors sustained on previous projects. Occurrences resulting in injuries also potentially increase the risk of schedule delays and the additional administrative burden necessary to deal with litigation that often follows personal injury occurrences.

Job-site managers generally concur that contractors who effectively meet technical specifications generally require less oversight than do those who are known to have failed to meet requirements on past jobs. Similarly, contractors with a high incidence of injuries can be expected to continue to experience losses at a similar rate in future work. Consequently, contractors with poor safety records require additional scrutiny and supervision. Further, they may put you and your employees at risk of injury as well. So, which contractor would you want on your job site? Which is the least expensive contractor in the long run? Most likely you would prefer to work with the contractor with a low rate of accidents.

II.   Contractor Screening

To assure yourself of having a safe job site, you must establish a process to screen out contractors with unacceptable safety practices. It is particularly important to select safe contractors for high-risk jobs such as construction, where loss exposures are the greatest. Contractors providing low risk services such as restocking food and drink machines, delivery of laundry, copy machine maintenance, and the like generally do not need to be subjected to the rigorous detail that would be applied to screening a piping contractor expected to work in a refinery environment. There are two approaches to screening contractors. The first is to develop a preferred bidders' list and the second is to evaluate the apparent low bidder after the bids have been opened.

Screening contractors to develop a preferred bidders' list has several advantages. Fewer approved bidders reduce the number of copies of bid proposal documentation that must be developed and sent out. Contractors whose past safety performance that would disqualify them from working on your project are saved the expense and unrealistic expectation of having a reasonable chance of competing. The development of a screened bidder's list enables an operation to begin to develop a strong safety program where the same contractors will be used with regularity.

The second approach is to evaluate the compliance of the apparent low bidder against the same criteria that would be used to develop a preferred bidders' list. A clearly defined and approved procedure must be in place to use this approach; otherwise it will be difficult to convince management after they have seen all the bids, to reject the lowest bid that may not be acceptable based on the technical safety selection criteria. A rationalization that seems to kick in is, "Well the difference is $50,000; for that price our safety guys can keep a closer eye on the job. It is only going to be for 9 months."

III.   Safety as a Technical Evaluation Criterion

Just as selection criteria are established to evaluate contractors' capability to produce a product to defined technical specifications or to demonstrate the financial robustness to remain in business throughout the duration of the contract, so should criteria be established to judge the contractor's ability to complete the work without personal injuries that would expose the host employer to subsequent litigation.

The effectiveness of the contractor's risk reduction and loss control practices should be the basis for contractor safety selection criteria. The principle being applied in this selection process is that organizations, and the individuals in them, will continue to behave as they have in the past. They are unlikely to change their work practices unless something significant forces them to do so. It is unlikely that a contractor with a history of high injury rates will all of a sudden stop having accidents simply because your contract states that you place a high degree of importance on safety. Appendix 8-1, Evaluation Criteria for Contractor Selection and Form 8-1, Contractor Safety Evaluation are provided as examples of the manner in which a number of organizations are defining their technical selection criteria and the format in which they are requesting the information.

Commonly used measures of contractor safety performance effectiveness are

1. Experience modification rating
2. Injury frequency and severity rates
3. Safety program evaluations and evaluation of the key personnel
4. OSHA citation history
5. References from others who have employed the contractor previously
6. Evaluation of the contractors' integration of safety into their work practices

Form 8.2 shows the contractor technical evaluation criteria to be used in conjunction with the contractor safety evaluation.

A. Experience Modification Rating

Criteria: Experience Modification Rate (EMR) of 1 or less.

Experience Modification Ratings (EMR) are calculated annually by each policyholder's insurance carrier. The calculation is based on past claims experience as compared to the average claims submitted by other policyholders in the industry in their respective states. The higher the policyholder's accidents experience and number of claims filed, the higher their EMR. The higher their EMR, the greater their premium paid to their insurance carrier.

The use of EMR rates as evaluation criteria is unbiased, as the contractor's own insurance company calculates the rate. An EMR of 1.0 indicates the company submits injury claims at a rate that is considered average for their industry group. A rating greater than 1.0 indicates the company files more injury claims than other similar organizations. A rating of less than 1.0 means the company has filed fewer injury claims than the average contractor in their industry.

It is becoming common practice to request that contractors submit their EMR rates for the past three years as a part of the evaluation criteria to assess the contractor's relative accident experience and to determine their accident experience trend. Since the objective is to work with contractors whose accident experience is at least as good as the average in their industry, the threshold acceptance criterion is generally an EMR of 1.0 or less. High-risk environments such as petrochemical plants generally set more demanding criteria and may only accept contractors with EMR rates of 0.7 or lower.

The comparison of contractor EMRs should be limited to those companies within the same industry group (SIC code) and only those from the same state. EMR rates for contractors with the same relative injury and claims experience may vary between states based on their interpretation of reporting requirements and the general work culture within each state. A contractor from Indiana with an EMR of 1 may have a different Lost Time Accident rate than a contractor from Michigan with an EMR of 1. The insurance industry recognizes this fact by varying the insurance rate from state to state. For example, in Indiana the workers' compensation rate for structural steel erection in 1996 was $16.96 while next door in Michigan it was $103.85 per $100 of payroll.

An important fact to bear in mind when evaluating EMR rates is that they can also be manipulated to some degree. An organization with a high deductible will file fewer claims than one with a lower deductible. In this situation it will appear that they have experienced fewer injuries. A tactic of devious firms with high EMRs is to dissolve and then re-incorporate under a new name. They then resume business with an EMR of 1 and continue with their unsafe work practices.

B. Injury Frequency and Severity Rates

Criteria: Accident Experience and Rates equal to or less than industry experience for similar industries as reported by the Bureau of Labor Statistics (BLS) or the National Safety Council.

Historical injury information provides a retrospective view of the level of safety practiced by a contractor. The evaluation criteria commonly used by most organizations in the U.S. are the following:

- fatalities
- injury cases resulting in lost workdays
- work days lost as the result of injuries
- injuries resulting in restricted work
- number of work days of restricted work
- cases requiring medical attention

1. Injury Case Rates

To equitably compare injury and illness experience, injury cases are converted into incidence rates that are based on man-hours worked. Although contractors are usually asked to provide their incident rates, it is often the case that they do not know how to calculate them.

To verify incident rates, it is common practice to request contractors to provide raw accident information numbers and the number of man-hours worked for each of the past three years to verify the rates quoted.

Incident Rate Calculation

Incidence rates are calculated using the following generally accepted formula:

$$\text{Incidence Rate} = \frac{\text{No. of incidents*} \times 200000}{\text{No. of hours worked}}$$

* Lost work day cases + cases involving days of restricted work activity.

116

Contractors should be asked to provide injury and illness data from all sites where work has been performed. If the contractors have affiliated businesses, e.g., hourly workers in one and white collar/management in another, that information should be provided as well. Contractors should have this information on hand as all companies with 10 or more employees are required to maintain an OSHA 200 log. This log is a record of illnesses and occupational injuries that have occurred during the past calendar year.

Clarification of Definitions

When asking contractors to provide their injury incident rates, there must be a clear definition of terms and equations used to calculate them.

a. Man-Hour Base

Some organizations may use 50,000, 100,000, or 1,000,000 man-hours instead of the BLS and OSHA base of 200,000 hours to calculate their incident rates. Clearly, the base hours used make a significant difference in the apparent loss rates.

b. Corporate Safety Philosophy

Another relevant inquiry is to determine the corporate philosophy of assigning injured employees to restricted or light-duty work. Some organizations return injured employees to work as quickly as possible after accidents, even if the employees' productivity is limited. Others prefer that injured employees take as much time as is legitimately needed to recuperate before returning to work. These two very different approaches have significant impact on frequency and severity rates.

Both physical absence from work and restricted work following an injury are considered lost time by OSHA. Organizations with aggressive injury management programs are increasingly encouraging employees to return to work as soon as they are able, even if in a modified duty capacity. It is their experience that employees return to full productivity sooner than employees who delay their return to work.

Further, the definition of time away from work must be understood, particularly in international construction environments. In some locations counting lost work days starts with the next regularly scheduled work day following the injury. Other locations do not begin measuring "lost time" until the employee has been off work for at least 3 work days. Adding to the potential confusion is the practice of some companies of switching employees' days off to minimize lost time reporting.

c. Head Count

The method by which contractors determine the number of "employee hours worked" used in loss-rate equations is also a relevant subject for inquiry. Does the number of man-hours used represent all employee hours, including those of

117

office workers? Office workers are exposed to fewer hazards; therefore, have fewer accidents. Including home office work hours, while perfectly legitimate, will reduce loss rates. It is not unreasonable to include home office hours as long as the basis of comparison is consistent between organizations. Consider requesting contractors to provide their man-hours in 2 segments, office and trade hours, to provide the information for an equitable basis of comparison.

d.   Types of Accidents

Ask contractors to provide a description of what caused their accidents and what they did to prevent the recurrence of those incidents on future jobs. Do they have a proactive program in place to continuously improve their safety program?

Is the contractor experiencing the same type of occurrences repeatedly? What is causing the accidents? Are injuries occurring repeatedly to the same work crew supervisors? If so, we certainly wouldn't want those particular supervisors on our job site if the contractor's experience as a whole meets our established evaluation criteria.

C. Safety Program Evaluations

Criteria: Integration of safety into work practices and guidelines for doing so.

A third evaluation measure is to analyze each prospective bidder's safety program and their compliance with safety regulations in the field. The evaluation of a contractor's safety program documentation provides useful insight into contractor awareness of hazards inherent in their work and in the environment in which they perform their work. The safety program should provide evidence of the necessary controls and accepted work practices to perform the work safely. If the program does not address this effectively, the organization may not have effectively analyzed their work or appropriately educated their employees.

Although the absence of a documented safety program does not necessarily mean that safety is not being integrated into work planning and execution, it may be an indication that there is not a uniform approach or standard by which safety is managed. This is not to suggest that every line manager must manage safety exactly the same way. However, generally, for a safety program to be effective there must be a common philosophy regarding safety implementation. Otherwise there exists the potential of conflict and error.

The evaluation of safety program documentation should verify that contractors are conducting regular work-site inspections and are scheduling specialized inspections as required. Inspection reports should be evaluated for consistency and frequency of inspections, thoroughness of the inspection, clarity of the report, appropriate classification of identified hazards, and timely and appropriate remedial actions. Specialized inspection records such as preventive maintenance and pre-use inspections should be checked to ensure their regular application.

Benchmarks against which to evaluate general corporate safety documentation are OSHA 29 CFR 1926 and the American National Standards

118

Institute's Construction and Demolition Standards (ANSI A10). Both of these references contain directions regarding what industry consensus has defined to be minimum criteria for effective safety programs.

Some organizations hire consultants to conduct annual program reviews. This brings in an unbiased set of eyes to evaluate the program. Another alternative is invite individuals from other work sites to conduct mutual inspections. If auditors identify problems that routine inspection reports have not addressed, the problem may be that the individuals conducting the routine inspections need more training in recognizing hazards and unsafe work practices. If the same observations are made repeatedly, there may be a problem with the process of controlling hazards resulting in the hazard recurring.

A safety program evaluation is clearly more time consuming than analyzing injury statistics, but is ultimately more revealing as it enables the reviewer to focus on the contractors' safety practices in a work environment relevant to the work to be done. The object here is determine if they manage safely and that it is part of their routine work practices as opposed to managing safety as an afterthought of their work process. Remember, the chosen contractor is going to be a partner in your business for some period of time and the last thing you want is to work with an organization that takes safety less seriously than you do, thereby potentially disrupting your project execution.

The Contractor Safety Evaluation form, found in Form 8.1, formalizes the request for information discussed previously. The rigor of analysis of this information should increase as the risk associated with a specialized service is required or where a long-term relationship is envisioned. This evaluation is particularly important for work in a high-risk environment such as refinery or chemical plants or where a long-term relationship with a maintenance contractor or general contractor to manage a project is being contemplated.

D. OSHA Citation History

Criteria: Absence of a negative trend of OSHA citations.

Another source of information regarding a contractor's past safety performance is the evaluation of this OSHA citation history. This information is often neglected out of concern for requesting confidential information. In fact, OSHA inspection records are public record and can be obtained directly from OSHA through the Freedom of Information Act if the prospective bidder is reluctant to provide this information.

Just as the absence of citations does not indicate outstanding safety performance, the presence of citations should not be an automatic basis for rejection. Where there are repeated citations for the same type of violation or regularly occurring citations that span some extended period of time, there may be cause for concern.

If a contractor has been the subject of regular OSHA scrutiny and citations, they may not be a contractor you want on your job site. The presence of that contractor may motivate OSHA representatives to consider visiting your site more frequently than they might have in the past.

CONTRACTOR SAFETY EVALUATION

Contractor: _____

Date: _____

1. INJURY/ILLNESS HISTORY

Provide the following injury and illness information for the past three calendar years.

	19 __	19 _	19 __
Number of lost workday cases	____	___	____
Number of lost workdays	____	___	____
Number of restricted workday cases	____	___	____
Total number of recordable cases (medical attention, sutures, foreign bodies in eye removal, rashes, etc.)	____	___	____
Employee hours worked each year	____	___	____
Number of fatalities	____	___	____

Incident Rates

OSHA Recordable Incident Rate   ____   ___   ____

$$\frac{\text{Number of Recordable Cases X 200,00}}{\text{Total Manhours Worked Per Year}} = \text{Recordable Incident Rate}$$

Example: $\frac{(5)\ Number\ of\ Recordable\ Incidents\ X\ (200,000)}{Total\ Manhours\ Worked\ (144,000)} = \frac{1,000,000}{144,000} = 6.9$

Lost Time Frequency Rate   ____   ___   ____

Lost Time Severity Rate   ____   ___   ____

2. WORKERS' COMPENSATION INFORMATION

Provide your Workers' Compensation Experience Modification Rate (EMR) for the past three calendar years. (NOTE: This information can be obtained from your Workers' Compensation Carrier.)

19___ EMR___, 19__ EMR____, 19_ EMR_____

Name and address of current Workers' Compensation Insurance carrier:
Name: _____
Address: _____
_____
Phone: _____

Form 8.1 Contractor safety evaluation form—example.

3. SAFETY PROGRAM INFORMATION

(If "provide documentation"(PD) block is checked when you received the form, submit documentation of applicable information.)

Do you have a written safety and health program?	Yes	No	PD
Do you have a written safety policy?	Yes	No	PD
Do you develop a Job Safety Analysis for work tasks?	Yes	No	PD
Do you have a written drug and alcohol policy?	Yes	No	PD
Do you have a written employee discipline policy?	Yes	No	PD
Do you have a written emergency plan?	Yes	No	PD
Do you have a written Hazard Communication program?	Yes	No	PD
Do you provide MSDSs for your employees?	Yes	No	PD
Do you have a formal accident investigation program?	Yes	No	PD
Do you complete and file written accident reports?	Yes	No	PD

If so, how often are accident reports and report summaries sent to the following:

	Never	Monthly	Quarterly	Annually
Field Superintendent	——	——	——	——
Vice President of Construction	——	——	——	——
President of Firm	——	——	——	——

Do you conduct new employee orientations?	Yes	No	PD

If you do, do they include instruction in the following?

a.	Head Protection	——	——	——
b.	Eye Protection	——	——	——
c.	Hearing Protection	——	——	——
c.	Respiratory Protection	——	——	——
d.	Fall Protection	——	——	——
e.	Scaffolding	——	——	——
f.	Housekeeping	——	——	——
g.	Fire Protection	——	——	——
h.	First-Aid Facilities	——	——	——
i.	Emergency Procedures— Rescue/Evacuation	——	——	——
j.	Hazardous Materials and Chemicals	——	——	——

Form 8.1 (continued)

k. Confined Space Hazards   ___   ___   ___

l. Signs, Barricades, Flagging   ___   ___   ___

m. Electrical Safety, Lockout/Tagout Procedures   ___   ___   ___

n. Rigging and Material Handling Safety   ___   ___   ___

o. Excavation Safety   ___   ___   ___

Do you have a program for newly hired or promoted foremen which includes the following subjects?

a. Methods of Safety Supervision   ___   ___   ___

b. Safety Meetings   ___   ___   ___

c. Emergency Procedures   ___   ___   ___

d. First-Aid Facilities   ___   ___   ___

e. Accident Investigation   ___   ___   ___

f. Fire Protection and Prevention   ___   ___   ___

g. New Worker Orientation   ___   ___   ___

h. Disciplining Workers   ___   ___   ___

Do you hold pre-job safety briefings ?   Yes   No   PD

Do you hold regular "toolbox" safety meetings ?   Yes   No   PD

If so, how often?

         Daily___Weekly___Monthly___Other___

Do you hold site safety meetings for field supervisors? Yes   No   PD

If so, how often are they held?

         Weekly___Bi-Weekly___Monthly___As Required___Never___

Do you conduct field safety inspections?   Yes   No   PD

If so, how often are they held?

         Daily___Weekly___Bi-Weekly___Monthly___As Required___

Identify by name and title the person within your firm directly responsible for the firm's Safety Program management:_____

     Phone: _____

Form 8.1 (continued)

While pursuing this line of evaluation, your corporate counsel may be able to produce records regarding personal injury litigation or property damage which may have been filed against the contractor. This will give you further indication of the type of safety-related problems the contractor may have had in the past that might be detrimental to your job progress.

E. References from Previous Employers

Criteria: Were previous employers satisfied with the contractor's safety performance?

Ask for references. Was the contractor responsive? What problems did they experience that you will have to deal with if that contractor should come to work with you?

How do the contractors, in fact, perform work on the job. Do they say they do one thing and then perform in a different manner?

F. Integration of Safety on Current Jobs

Criteria: Demonstrated effectiveness of integrating safety into current work practices on current jobs.

The most effective means of evaluating a contractor's ability to work safely is to visit a job site where they are working to see how they perform work. If you know you are going to require cranes on your project, take a look at the cranes the contractor is using. What condition are they in? How are they being managed? Are they equipped with load moment and anti-two blocking devices? The condition of their work area and how they manage their work will give you some insight into the corporate culture regarding the standard of care they apply to their work.

Interview the prospective contractor. Does he have a documented safety program? If so, does the program address the hazards to which workers are exposed and which comply with applicable regulations. Can the contractor manager you are visiting produce his documented safety program? The prospective project manager's attitude regarding safety should also be taken into consideration. There are only a limited number of "A" players available and you want to make sure that even if the corporate office gives you the indication that they are the right organization for you, the individual sent to your job site must have the right attitude as well.

Therefore, the criteria should be applied to both the contractor and the performance of contractor's managers who will directly influence the work on the job site. What better means of improving the odds of good safety performance on a project than the selection of organizations and individuals with proven track records?

IV.   Request for Bid and the Pre-Bid Meeting

After establishing selection criteria and developing the Request for Bid (RFB) documentation, prospective bidders should be invited to a Pre-Bid Meeting. Prospective bidders should be provided with an opportunity to review the RFB requirements, ask questions and obtain clarifications as directly as possible.

Clear and concise definitions of safety requirements in bid invitation documents are important elements in effective contractor safety management. Prospective contractors must understand the project's safety requirements before they can develop a realistic bid. Implied requirements or requirements that are not clearly stated in the bid documents may become the source of conflict later. Many conflicts that develop during the contracted work process can be traced to misunderstandings and miscommunications regarding requirements and standards that the job-site manager considers relevant.

Where a preferred bidders' list has not been developed and the lowest apparent bidder is to be evaluated, the safety evaluation criteria to be used should be defined prior to the start of the contract bidding process. A statement to the effect that safety will be included as technical evaluation criteria should be included in the RFB. It is only fair that prospective contractors be made aware that their past safety performance will be evaluated, prior to their investing time and effort in bid preparation. Contractors who recognize that they have a poor safety record or a weak safety program may opt not to participate in the bidding process with the knowledge that there exists a real possibility that they may be rejected based on their poor safety record. With this knowledge, if in the event the lowest apparent bidder is not awarded the contract, there will be no basis for the low bidder to contest the bid evaluation by claiming he was not aware of the safety selection criteria.

To ensure that prospective contractors have the information they require to develop estimates to address job-site hazards, some organizations go as far as including their phase analysis evaluation as an addendum to their RFB. It may not be necessary to include the entire analysis, but the RFB should address at least the following items:

A.   Work to be done
B.   Identified hazards and work restrictions
C.   Work permits and license requirements
D.   Contractor safety qualification requirements
E.   Special job-site safety program requirements
F.   Contract management arrangements
G.   Orientation and training requirements
H.   Audits of contractor performance
I.   On-site control of work

J.   Code compliance requirements

K. Record-keeping and reporting requirements

L. Contract termination/completion criteria

A. Work To Be Done

The work to be done should be described in sufficient detail to enable prospective contractors to understand the content of the work required as well as the work environment.

B. Identified Hazards and Work Restrictions

Unusual hazards and potential restrictions should be brought to the contractor's attention in the RFB and repeated during the Pre-Bid meeting. It may sometimes be difficult for prospective contractors to get an accurate picture of the scope of the work in a particular project by just reading the contract documents. A good way of communicating this information is to include a work site visit in the Pre-Bid meeting agenda. This provides the prospective bidders with an opportunity to see the job site and ask questions relevant to their bid development. Make sure that all requests for clarification are also submitted in writing. Then provide written answers to all the prospective bidders.

C. Work Permits and Licenses

Work permits and licenses required by the site or local code should be identified. The methods by which they will be issued and controlled should also be addressed. Some areas to be considered are

- On-site entry and security
- Vehicle passes
- Confined-space entry
- Excavation
- Lockout/Tagout requirements
- Working hot (electrical)
- Explosive devices
- Hazardous materials handling and transportation
- Hazardous waste disposal
- Radiation source management
- Environmental permits

The RFB document should indicate that the successful bidder and his subcontractors will be responsible for the enforcement of safe work practices outlined in the permits and licenses, as well as all other site or local code requirements that apply to the work under their control.

D. Contractor Qualifications and Requirements

Contractor qualifications are those criteria the selection team will use to evaluate the prospective contractors and their bids.

E. Special Job-Site Safety Program Requirements

The job-site manager must define all site-specific safety requirements. These might include specific types of tools, restrictions on certain work practices and special lockout procedures, or site-specific fall protection requirements. The contractor and his subcontractors should provide detailed documentation that demonstrates that their safety programs meet the job-site manager's requirements.

F. Contract Management Arrangements

The RFB document should define the job-site organization hierarchy and the key contact points. Critical questions which should be addressed are

- What is the job-site organization structure?

- Who within the host's organization is responsible for contract coordination and administration?

- Who will address safety and health issues?

- What technical support, if any, will be provided?

- How will work delays, challenges, accidents and similar events be reported to the job-site coordinator?

- What information must the contractor report to the job-site coordinator during the life of the contract and to whom should the reports be made?

- What anticipated hazards will be present on the job site and how will the host control them?

G. Orientation and Training Requirements

The RFB should identify specific training requirements deemed necessary by the job-site manager for successful completion of the work. The contractor should be required to demonstrate to the satisfaction of the job-site manager that its employees and subcontractors have completed the site orientation and job-specific training, as well as any other qualifications specified by the job-site manager.

Some host organizations prefer to provide and manage certain orientation and training requirements they feel are particularly important prior to granting contractor employees permission to be on the work site rather than relying on the contractor to communicate this information.

H. Audits of Contractor Performance

The RFB should specify that the job-site manager will conduct periodic assessments of the contractor's safety and health program implementation during the contract period. A copy of the audit protocol and performance requirements should be available if the contractor would like to see what would be evaluated. The job-site manager should carefully consider the practicality of each type of measurement taking into consideration the length and risk level of each individual contract.

I.   On-Site Control of Work

Contractors should be expected to have a process in place to monitor for unsafe work practices. Appropriately trained individuals must do this. Specific requirements are now in place that define the individuals who are recognized as being able to competently inspect specialized work such as excavations and scaffolding. The contractor should be asked to explain the basis upon which these individuals have been determined to be competent to inspect the work. The contractors should also be aware that they will be expected to provide for direct supervision of employees at all times. Ideally, the contractor should provide its senior managers, supervisors, and on-site safety coordinator with training in hazard recognition, and safety and health management.

J.   Code Requirements

The RFB documents should identify specific code requirements, relevant regulations including licensing of individuals and certification of equipment.

K.   Record Keeping and Reporting Requirements

All records and reports required by the job-site manager pertaining to the contractor's safety program should be identified. Such records generally should include

- Weekly Personal Injury Reports and Man-hour Reports
- Records of personnel training
- Documentation and follow-up of identified safety problems
- Inspection Reports.
- Accident/Incident Investigation Reports
- Equipment Inspections
- Safety Meetings

Standard forms to be used for records and reports may be specified in the RFB; however, the major concern of the job-site manager should not be whose form is to be used, but that the critical information on each issue is provided.

L.   Contract Termination/Completion

The RFB documents should clearly define contract penalties and termination procedures in case of substandard safety performance. The documents should also specify when and how disciplinary measures will be imposed. Consideration should also be given to defining what will be required to make the determination that the contract is judged to be satisfactorily completed.

V.   Summary

Preference for contractors with good safety experience will improve the probability of completing a project with a low injury incidence rate and the absence of major mishaps. If the selection of preferred bidders is not an option, then safety should be an integral aspect of the contract technical selection criteria.

Since construction is considered to be a potentially high-risk environment, contractor selection is being increasingly advocated in consensus standards such as ANSI[2] and recently in the OSHA Process Safety Management standard (PSM).[3]

A detailed job plan which identifies risks inherent in a project is critical to the successful management of contractor safety and health. If the "work to be done" or the expected performance standards are not set out clearly and completely prior to the bid request, prospective contractors may submit inadequate bids.

Get your legal counsel involved in the development of your contract language as it relates to safety requirements to make sure it is clearly defined. Then stick to your guns. The true test of senior management's understanding of the value of safety and supporting it is their being able to pass over the low bidder with a poor safety record in favor of the second or even third higher bidder able to meet the safety criteria requirements.

Thus, the value of selecting safety conscious contractors includes less exposure to injuries and insurance claims, greater control of hazardous conditions in the workplace, increased productivity, improved morale, and decreased liability to third-party lawsuits and contractual disputes.

Enlightened owners and project constructors will soon no longer accept generic safety manuals as evidence of a corporate safety program. The more proactive organizations are insisting on job-specific safety documentation. Project managers wish to see evidence of safety documentation directly related to the work to be performed on the job at hand.

---

[2]   Basic Elements of a Program to Provide a Safe and Healthful Work Environment, ANSI A10.38-1991.
   Safety and Health Program Requirements for Multi-Employer Projects, ANSI A10.33-1992.

[3]   Process Safety Management of Highly Hazardous Chemicals, OSHA Standard 29 CFR 1910.119 & 29 CFR 1926.64.

MEMORANDUM

Date:
To: Procurement Services
From: **Your Name Here**   Project Manager

Subject: CONTRACTOR SAFETY EVALUATION CRITERIA

Contractor:

Work/Project:

The following is the safety evaluation criteria which apply to the aforementioned contracted work.

Topic   **Criteria**

1. Experience Modification Rate   1 or less

2. Lost Time Case Index Rate   Industry average for SIC

    Lost Days (Frequency) Rate

3. Accident Information to Management   At least quarterly

4. Safety Plan and Job Safety Analysis   Yes

5. Hazard Communication Plan   Yes

6. Workers' Compensation Carrier Info.   Must be completed

7. Designated Safety Coordinator   Must be identified

8. Safety Training   Quarterly Toolbox Meeting

9. Safety Inspection   Bi-weekly

10. Employee Safety Orientation Program   Yes

Attachments: Contractor Qualification Information Form
                         Contractor Information
cc: Project Manager

Form 8.2 Contractor safety evaluation criteria

APPENDIX 8.1

EXAMPLE EVALUATION CRITERIA
FOR CONTRACTOR SELECTION[4]

Proposals for the fabrication and installation of the Nitrogen Distribution System will be evaluated by consideration of technical and cost/price factors. Job-site manager shall conduct technical evaluations based on the Technical Criteria listed in Paragraph A below. Criteria 1 through 5 are of equal importance and will be scored on a "Go/No Go" basis. Only those offers scored "Go" for each criterion will be considered <u>Technically Acceptable</u> and therefore eligible for award. Price will be evaluated in accordance with "B. Price and Price-Related Criteria."

A. Technical Criteria

Criterion 1 — Safety
The offeror must demonstrate the existence, within the offeror's organization and that of any selected subcontractor, of a satisfactory safety record which, in the job-site manager's view, *is sufficient to successfully complete this project.*

NOTE: *To satisfy this requirement, the offeror must provide documentation from the offeror's and any subcontractor's Workers' Compensation Carrier which shall reflect a Workers' Compensation Experience Modification Rate (EMR) of "1" or less and meet the selection criteria defined in Figure 8-1.*

Criterion 2 — Past Performance
The offeror must demonstrate the successful performance of at least two projects that, in the job-site manager's view, are substantially similar to the scope of this requirement within the past three years.

NOTE: *To satisfy this requirement, the offeror must provide details of each successfully completed project or individual product grouping, valued in excess of $100,000, to include the following information:*

1. Project description
2. Dollar value
3. Inclusive dates of performance
4. Clients who may be contacted as references to include name, address, and telephone number
5. Performance records regarding delivery within schedules and quality of performance
6. Explanation of the relevance of cited projects to this effort

---

[4] **Note:** This is an example of the information that might be included in an invitation to bid. It gives notice to the prospective bidders of the safety requirements to which they will be held.

Criterion 3 – Welding Capabilities

The offeror must demonstrate the availability, within the offeror's organization or that of any selected subcontractor, of a welding program that, in the job-site manager's view, *is sufficient to successfully complete this project.*

NOTE: *To satisfy this requirement, the offeror must provide copies of the offeror's or subcontractor's welder qualification program and transfer line construction welding procedures.*

Criterion 4 – Cryogenic Cleaning Capabilities

The offeror must demonstrate the availability, within the offeror's organization or that of any selected subcontractor, of a cryogenic cleaning program that, in the job-site manager's view, is sufficient to successfully complete this project.

NOTE: *To satisfy this requirement, the offeror must provide copies of the offeror's or subcontractor's cryogenic cleaning procedures.*

Criterion 5 – Quality Control

The offeror must demonstrate the availability, within the offeror's organization or that of any selected subcontractor, of a Quality Control program that, in the job-site manager's view, *is sufficient to successfully complete this project.*

NOTE: *To satisfy this requirement, the offeror must provide copies of the offeror's or subcontractor's Quality Assurance Manuals and/or Plans which shall detail Quality Assurance Procedures to be applied to this project which substantially incorporate ANSI/ASQC-C1, "Specification of General Requirements for a Quality Program."*

B. Price and Price-Related Criteria.

Bid prices price will not be numerically scored or adjectivally rated (some organizations place weighting factors on bids based on how much they deviate from the organization's original estimate). However, price and price-related factors specified in the solicitation are considered more important than the technical criteria listed above in the job-site manager's overall evaluation and will be the controlling factor for award when considering all technically acceptable offers. Technical acceptability is defined as any offer which was rated "Go" on all technical criteria listed in Paragraph A above.

Note: *This criterion was established for the installation of a nitrogen supply line for a laboratory facility.*

Chapter 9

Preparing for Contractor Mobilization

I.   Introduction

Your procurement department has advised you that the apparent low bidder for your project's site preparation has been identified. It is now time to prepare for the contractor's mobilization.

First, the apparent low bidder's safety documentation must be reviewed to establish that they have taken note of the hazards relevant to the site and that their site-specific safety program addresses the manner in which they will control them. Next, you must assure that your own project safety program documentation, project information, and safety files are in place. Finally, has the site layout been defined and the requisite infrastructure support required for the project been coordinated? Once this has been done you will be in the position to issue the contractor a "Notice to Proceed."

II.   Safety Program Documentation

Often the weak link in contractor safety programs is the contractor's failure to recognize hazards specific to the site where they are to work and therefore, they have failed to develop an effective means of controlling those hazards. Although it is the responsibility of the contractor to develop a sufficiently detailed safety analysis and site-specific safety plan to deal with those hazards, it is in the best interest of the job-site manager to ensure that the contractor has done an effective job in this regard. Unfortunately, it is frequently the case that the review of the contractor's plan and monitoring of its implementation are overlooked by the job-site manager in his single-minded focus on getting the work started.

A. Site-Specific Safety Plan

To ensure each contractor is in a position to comply with the contract safety requirements, the job-site manager or a designated field engineer should review the contractor's safety program to verify that it addresses the work defined in the contract. Enlightened owners and project constructors will no longer accept generic safety manuals from their contractors as evidence of a safety program. Host employers with proactive safety programs insist that contractors have job-specific safety plans that are directly applicable to the work to be performed.

The time to review the contractor's site-specific safety plan and to ensure that it is complete, is prior to their mobilization. Since contractors invited to bid should have met your technical safety screening requirements, in the worst case the selected contractor should only need to make minor changes to his safety program to meet specific project safety requirements. Where the contractor is required to modify his procedures or develop new ones to satisfy contract requirements, he should be able to do so relatively easily. The job-site safety program and information incorporated into the contract by reference should provide ready reference material for the contractor to develop missing procedures.

At this point it is likely that a warning flag is waving in the back of your mind. Your concern is likely to be that by exercising this kind of control you will be liable in tort if a contractor's employee is injured. Also, close oversight of a contractor's safety conduct can increase one's OSHA liability. Case law developed by the Review Commission and the courts indicates that the more involved with contractor safety that a host or higher-tier employer becomes, the more vulnerable to citation that employer becomes if the contractor commits a violation. In fact, some case law appears to require the very involvement that gives rise to additional liability. However, the fact is that in addition to controlling the work site, the host employer is the most knowledgeable about the hazards therein. The host employer's best defense against contractor losses is the assurance that the contractor understands the hazards and can demonstrate that understanding through documentation of how he proposes to deal with each hazard.

By reviewing the contractor's documentation there will remain little doubt whether or not the contractor really understands the hazards and how to deal with them. That is the host employer's best defense.

B. Contractor Information Sheet

Each contractor must submit a list of their key personnel, their designated safety representative and competent persons along with their respective phone numbers. As the host employer, you want to know who the players are and how to reach them if necessary.

C. Chemical Inventory and Material Safety Data Sheets

Control of hazardous substances and work by-products is the contractor's responsibility. This includes noise, dust, toxic compounds, radiation, welding flash, etc. A complete chemical inventory and Material Safety Data Sheet (MSDS) for every chemical product to be brought on site should be readily available and a copy submitted to the job-site manager by the contractor prior to starting work. The contractor must have a copy of MSDSs for all chemical compounds that are on site at their on-site "office." Additionally, an MSDS should be attached to the relevant JSA posted at the work site. You must ask for this information up front or else the contractor will have one excuse after another of why he has not yet produced it. When and if needed, it may not be available. The medical staff in some hospitals are reluctant to treat chemical injuries without specific information on the chemical source. The health of the employee may be at risk if the proper MSDS cannot be produced promptly.

D. Job Safety Analysis

Contractors should also be required to forward a copy of their JSAs for each work activity they are expected to perform to the job-site manager as a condition of being granted a notice to proceed. The job-site manager or his field engineer should then review them. The object of the review is to assess the effectiveness of the proposed hazard controls. This review offers the additional value of providing an insight into the procedures the contractor intends to follow in performing his work. This information will also enable the field engineer to identify potential conflicts with concurrent activities being performed by other contractors. The exception to the requirement that all JSAs be submitted for review and endorsed before the contractor is granted a "Notice to Proceed" would be in the case of protracted jobs. It may be agreed in those cases that JSAs could be submitted for review prior to the start of each predefined work phase.

If a contractor is not familiar with JSAs, it is reasonable to provide examples produced by others or even examples such as are produced by the many organizations that sell them. The important point is that the contractor must declare that the document is one that he has produced and that it is the manner in which he proposes to perform his work.

It is also important that the JSA be signed off by each and every employee expected to perform the work. Form 9.3 is the typical format for JSAs. Some organizations include a short list of site-specific rules in the JSA that generally apply to all work as a reminder to the contractors. It is of little importance if the document is typed or handwritten. The fact that it has been prepared, communicated to each and every contractor employee, and that the contractor is expected to perform to those requirements is clear evidence that they knew or should have known about the job-site hazards and that they had been briefed on how to perform the work safely.

Contractor Safety Requirements

Before a contractor is permitted to start work, they must participate in a job-site orientation and submit the following information:

- A site-specific company safety and health program,
- A completed Contractor Information Sheet (names and telephone numbers of key management, safety, and competent persons),
- A chemical inventory and MSDSs for all chemicals on site, and
- Job Safety Analysis (JSA).

In addition, each contractor is required to

- Conduct job-specific training for their workers based on the content of their JSAs,
- Provide and document HAZCOM training to all employees,
- Provide training in PPE, fall protection, confined entry, and other safety training as specific to their work,
- Have evidence that employees have completed a negative drug screening,
- Provide ground-fault protection for all non-permanent wiring,
- Post emergency telephone numbers by telephones, and
- Define how medical services will be provided.

III.   Job-site Manager's Preparation For Mobilization

The job-site manager's site-specific safety program documentation and information required for the effective execution of the work project should be in place before the first contractor mobilizes to the project site. The Pre-Job Planning Checklist (Form 9.1) provided at the end of the chapter is an example of the considerations that should be included in preparation for contractor mobilization and the sort of documentation that should be in place at the start of the work project. Clearly, the checklist developed for each project will reflect the considerations specific to it. A grassroots refinery will include different considerations than will those of a steel mill upgrade. The development of the checklist will be based on notes and information gathered during the engineering and procurement phases of the project. The checklist serves to ensure that nothing is overlooked in preparation for contractor mobilization.

A. Project Safety Program & Manuals

Site-specific safety documents, permits, forms, safety-related information and required posters should be available in the project office. Files should be established prior to the start of the project to ensure that no safety-related information determined to be important is overlooked. Clearly, the contents of the file will be influenced by the size and duration of the work to be done.

> Safety and Health Contacts
> Safety Meeting Minutes
> Personnel Qualifications
> Safety Training Records
> Inspection Records and Related Hazard Abatement Information
> Job Safety Analysis Forms
> Material Safety Data Sheets
> Accident Summary Log
> Injury and Illness Reports
> Toolbox Talk Minutes

1. Safety and Health Contacts (Project Organizational Chart)

An organizational chart with the names, titles, and phone numbers of key project personnel is an important reference document for contractors new to the facility. This information will enable them to be more responsive and less dependent on the job-site manager or field engineer for direction. Individuals responsible for issuing permits or who are required to conduct inspections prior to the approval to start permitted work should also be included in this list of contacts.

2. Safety and Health Program

The site safety and health program should be available as a practical guide and interpretation of contractual safety requirements. The information in this document should be phrased in terms that make sense to field personnel. It should be available for reference on the arrival of the first contractor employee. A well-organized owner may already have such a document in place; otherwise the job-site manager will have to produce one.

3. Posters and Permits

OSHA addresses posting notices, including the OSHA poster, in 29 CFR 1903.2. It addresses information required in states covered by Federal OSHA and for establishments operated in State-Plan states.

Job Safety and Health Protection posters such as the OSHA 2203 poster are required to be posted in a conspicuous location where notices to employees are customarily posted for each office, shop, and job site. This poster can be found on the Internet at (http://www.osha.gov:80/oshpubs/poster.html).

Injury/Illness Reports — OSHA 101, must be completed to record details of each work-related illness or injury and must be completed within six working days of knowledge of the incident. State Industrial Commission Reports

(Employer's First Report of Injury) are acceptable in lieu of OSHA 101 forms. These reports are not sent to OSHA, but must be retained by the employer for five (5) years.

If an "on-the-job" accident results in the death of an employee, or the admission of five (5) or more employees to a hospital, the Corporate Safety Director must phone the nearest OSHA office and provide a detailed report within 48 hours of the accident.

OSHA 200, Log and Summary of Occupational Injuries and Illnesses is required for each job-site and maintained on a calendar year basis. OSHA requires that each "Recordable" occupational injury, and all illnesses, be recorded on the log within six (6) working days after being reported to the employer. OSHA definitions of "Recordable" and instructions for completion of the OSHA 200 are on the reverse side of the form.

The OSHA No. 200 and OSHA No. 101 forms are to be maintained at the workplace or by the facility owner for five years and must be available for inspection by the Occupational Safety and Health Administration (OSHA), Health and Human Services (HHS), Bureau of Labor Statistics (BLS), or the designated state agency.

In addition to OSHA-driven requirements for forms and posters, the facility owner where the contracted work is to be performed may have specific poster and permit requirements. The following are examples of the types of site-specific forms and permit requirements in place at most established facilities:

Accident Investigation Forms
Excavation Permits
Jobsite Safety Inspection Forms
Confined Space Entry
Fire Protection System Impairment
Movable Structure Siting
Notification of Demolition and Removal (asbestos)
Open Flame and Spark-Producing Operations
Work Entry Clearance
Working Hot Electrically
Building Entry Permits

B. Site Layout and Facility Preparation

Site layout and space allocations should be included in the preparation for contractor mobilization. A well-thought-out site layout will reduce the potential for conflicts between ongoing facility operations and those of the contractors. Consideration should be given to such issues as security, owner operations, and safety of the public. To minimize the potential for misunderstandings and coordination conflicts the facility owner should be asked to review the proposed job-site layout for compatibility with ongoing operations.

1. Site Access Control

Security and site access control should be given detailed attention. Where possible on large projects contractors should be provided with their own access gates. If, by chance, both union and non-union workforces are being employed at the same location, each should be provided with their own job-site access gate. Where possible, contractor access to their work sites should not traverse areas where owner production activities are taking place.

2. Traffic Control

A traffic control plan with consideration for traffic leading to and within the confines of the project site is an essential consideration. The job-site manager should provide for adequate turning room and good visibility for vehicle drivers at job-site entry and exit points in the site layout plan. Set out clear vehicle routes across the job-site avoiding sharp bends, blind corners, and narrow gaps. Protect any temporary structures such as scaffolds that might be damaged or made unsafe if struck by a vehicle. Protect excavations and areas alongside open water that vehicles must pass closely. Site traffic patterns should be defined to minimize the need for construction vehicles and equipment to cross or come close to pedestrian paths.

Where possible, clearly delineated sidewalks or footpaths should be established. This is particularly important where the vehicles must come close to public pedestrians. Pedestrian movement around construction activities should be minimized where possible. All general, public foot traffic parallel to construction traffic must be protected and where possible, barrier systems installed. Normal vehicle curbing is not usually an adequate substitute for barriers where the general public is involved. Where pedestrians must cross the path of construction vehicles, clear cross-walks should be delineated and traffic rules enforced.

Workers are at a greater risk of dying from traffic-related accidents than from any other job-related hazard, according to the National Institute for Occupational Safety and Health (NIOSH). Traffic-related crashes were the leading cause of fatalities, accounting for 20% of the on-the-job deaths while driving, riding, or working around motor vehicles in the U.S. from 1980 to 1992. 60% of workers killed in vehicle accidents were not wearing seat belts. A requirement to wear seat belts while on the job-site is much like the requirement to wear hard hats. They both create an additional sense of awareness regarding safety.

3. Equipment Location and Storage Areas

Hazards created by equipment such as generators, air compressors, and welding machines include noise, dust, and toxic fumes and should be considered when determining their placement. Prevailing winds and their effect on directing toxic fumes and dust should be the first of the elements considered. Isolation of large noise producers in remote locations or the installation of noise deflectors and sound barriers should also be considered when in the vicinity of densely occupied work areas.

4. Miscellaneous Services

Trash disposal, servicing sanitary facilities, and provision of potable water should all be established and addressed during the work release meeting. Arrangements should be made for these services to be provided on a schedule that will allow for an orderly workplace, minimizing work disruption and excessive accumulations of waste material.

C. Material and Equipment

1. Personal Equipment and Instrumentation

The contractor is generally required to provide all safety materials and personal protective equipment for their site safety program. The job-site manager should be prepared to monitor the contractor's work activities to ensure appropriate equipment that meets established standards is being provided.

Personal Protective Equipment

Safety hats	Chemical and acid
Safety shoes	Respirators and filters
Rubber boots	Dust
Gloves	Toxic chemicals filters
Abrasion and puncture	Paint spray
Welding and burning	Full face shields
Electrical	Welders hoods
Acid and chemical	Aprons
Hearing protection	Safety harnesses
Plugs and muffs	Life jackets
Eye protection - glasses/goggles	Sandblast helmets
Impact	Sweat bands
Burning and welding	Goggle cleaning stations
Dust	

On remote projects it is often in the interest of the job-site manager to consider maintaining a stock of personal protective equipment that can be drawn out by the contractors when needed in a pinch. The list of equipment above is offered only as an indication of the variety of equipment that may be considered. Fortunately, there are a growing number of organizations that supply safety equipment that can be tapped to provide this service.

2. General Use Equipment

On larger projects it is often in the interest of the job-site manager to have a stock of general use material and equipment. Clearly, the variety of material will once again depend on the activities being performed. Although it is generally the responsibility of subcontractors to provide this material, having these items available can expedite the execution of work.

Barricade posts, rope and tape	Trash containers
Traffic signs	Rags
Red cloth for hazard marking	Explosimeter
Fire retardant tarpaulins	Gas testers
Rubber dielectric blankets	Carbon monoxide indicators
Exhaust system or blowers	Radiation dosimeters
Flammable liquid safety cans	Safety bulletin boards
Welding curtains and shields	Noise measuring devices

By this point it is apparent that the list of issues which should be addressed in the preparation for contractor mobilization is not a trivial consideration. The fluidity with which contractors can be integrated into the work process depends to a great extent on having identified and prepared for all significant eventualities.

3. Communication

How the job-site manager proposes to communicate with the contractors during routine activities, as well as in emergency situations, should be established. On large jobs with numerous contractors, it might be in the best interest of the job-site manager to provide radios. Where circumstances permit, a site-wide public address system might be established early in the project's life cycle for emergency communication to all workers.

4. Fire Protection & Fire Fighting Equipment

The provision of fire protection services depends on the project, its location, size, fire hazards, and other factors. On large remote projects, a dedicated fire department or volunteer organization might be established. In municipal environments, arrangements can be made with the local Fire Departments for support. The facility owner may define the level of coverage required and where that support will be obtained.

5. First Aid & Medical Support

On large projects with "Wrap-Up" insurance, a nurse and sometimes a doctor are on site to provide immediate medical attention. On projects of smaller scope, a Fire Department may agree to respond and provide paramedic support. At the very least the contractors should define how they propose to provide medical support to their employees in the event of both minor and more serious personal injuries. The more prompt the medical service, the fewer the complications and greater the likelihood of the worker returning to work without a protracted absence from the job site. Employees who are not at work are obviously not productive workers. If the job-site manager is not able to provide for immediate medical response, the contractor should be required to have a first aid- and CPR-trained first responder on his work crew.

6. *Emergency Preparedness and Support*
Given the importance of emergency preparedness, Chapter 14 is dedicated to this subject.

D. Education and Training

1. *Employee Orientation*
Every contractor, direct hire employee, and visitor who will not have a full-time construction safety-trained escort should attend an initial work-site orientation prior to being given site access. At a minimum the orientation should address:

- Project/safety organization
- Site safety rules, procedures, and safety signs
- Alcohol and substance abuse policy
- Hazard communication program
- Employee rights and responsibilities
- General safety procedures and site-specific hazards
- Personal protective equipment requirements
- Accident reporting and the location of first aid and medical facilities
- Safe work permit requirements
- Vehicle use and parking restrictions
- Eating, drinking, and smoking restrictions
- Disciplinary procedures
- Sanitation and housekeeping
- Waste disposal and dumpster usage
- Emergency response guidelines
- Fire alarms/safety showers/eye wash stations
- Location and access to the approved project safety plan

An orientation checklist such as that shown in Form 9.2 should be attached to the roster and dated and signed by all the individuals who attend each orientation. This training record will become an important document during an OSHA or host employer safety audit.

It is a common practice to give each attendee a dated and signed card certifying completion of orientation. This is the employee's evidence of having attended the orientation if job-site access identification cards are not issued. On larger projects a database is maintained with the social security number and date of each individual attending the safety orientation. This information provides construction workers who might change employers with evidence that he has attended the training within the past year and saves the project the cost of having the worker attend it again.

2. *Training Records*

Training files should be maintained for all required training. This record should be available upon request at each hierarchical level within the project for review by the project manager, project engineers, and other personnel with assigned oversight responsibilities.

3. *Job-Specific Training*

In addition to the general job-site orientation, the job-site manager should have a process in place to ensure that each worker receives job-specific training. This is training conducted by their direct employer, which addresses the specific safety considerations related to the work to be performed by that contractor.

E. Equipment & Tool Inspection

Contractors are required to provide the tools and equipment necessary to complete their work. It is in the interest of the job-site manager to be assured that the equipment brought to the project is in good working order and that the equipment is in compliance with OSHA and ANSI requirements. To this end, a tool and equipment inspection process should be established.

My experience has been that at the start of each project less than desirable heavy equipment is sent to the job-site by rental agencies. Crane hydraulic systems leak, hardware such as anti-two blocking devices is missing, and rigging is in poor condition. Earth-moving equipment such as bulldozers arrive without rollover protection. The risk of using less than totally reliable and outfitted equipment is that the equipment may break down resulting in project delays. Through rigid adherence to high criteria for equipment appearance and maintenance, soon only good equipment will show up at the gate. Less desirable equipment will be sent elsewhere.

Personal tools should be subject to a similar inspection process. Although contractors are responsible for ensuring that their workers' tools and equipment are in good working order, it has also been my experience that many do not fulfill this responsibility. The number of defective tools found during tool inspections will provide an indication of the level of oversight that is likely to be required of the contractor. Some things that should be checked and judged for acceptability are

- Guards on exposed moving parts of tools and equipment.
- Compatibility of replacement of expendable tool parts such as grinding wheel and saw blade design speeds.
- Electrical insulation or grounding on electrical tools and machines.
- Fail-safe control mechanisms on powered tools.
- Tool bit retainers on powered tools such as chipping guns and paving breakers.

F. Communication & Publicity

1. Communications

Contractors should be required to report the status of their site safety program to the job-site manager at least monthly. The report should include safety statistics, significant safety activities, safety problems and progress in resolving them. In addition to formal reporting, each job-site manager's project team member should establish lines of communication with the contractors that will enable both parties to resolve problems expeditiously and be cognizant of the safety activities in their areas of responsibility. In addition, there should be an exchange of safety information among the project team members. A good way to accomplish this is to hold regular safety meetings with the contractors and the project team members. Feedback from safety meetings provides an indication of interest and management commitment to the safety program.

2. Reporting of Significant Occurrences

Frequency and severity of injuries are the accepted quantitative measures of safety performance. It is necessary, therefore, that the recording of injuries be done accurately, particularly if there are incentive schemes related to contractor safety performance. The project team review should ensure that the forms used by contractors provide the desired information. The contractor must investigate all accidents and a copy of the reports submitted to the job-site manager. Where possible, the job-site manager should designate an observer to participate in the contractor's accident investigation.

3. Safety Program Awareness

A safety publicity program to promote safety throughout the construction period should be contemplated. Contractors need regular reinforcement from the owner or job-site manager that safety is important. This safety awareness program may take the form of an incentive program or identification of the Contractor and Contractor-Worker-of-the-Month types of programs.

IV.   Summary

A great deal of "up-front" work is required to prepare the job-site for contractor mobilization. The better prepared the site is for this event the fewer the disruptions and stumbling blocks that will be encountered. It should be the job-site manager's objective to get each contractor onto and then off the job-site as fast as possible and with little disruption to other ongoing activities. Through detailed planning, coordination and communication with each contractor this can be achieved.

Form 9.1 Pre-Job Planning Checklist

	ACTION BY	COMPLETED
I. Contractor Safety Documentation: Confirm that the following documents have been developed, reviewed and approved by appropriate parties.		
1. Site-Specific Safety Plan		
2. Job Safety Analysis of Each Work Step		
3. Material Safety Data Sheets for All Hazardous Materials to be Used on Site		
II. Safety Baseline Assessment: Confirm that completed and approved safety assessment documents are on file.		
1. Baseline Safety Assessment Complete		
2. Environmental Assessment		
3. Phase Analysis Complete		
III. Project Safety Program & Manuals: Confirm the following project plans have been developed, reviewed and approved by appropriate parties and are in place.		
1. Safety & Health Contacts (Org. Chart)		
2. Environment Safety & Health Manual		
a. Heath & Safety Standards		
b. Permit Forms and Posters		
i. Poster (OSHA 2203)		
ii. OSHA 200 Log		
iii. Injury/Illness Reports (OSHA 101)		
c. Emergency Management Plan		
i. Evacuation, Assembly & Personnel Accounting Procedures.		
ii. Medical Emergency Procedures		
iii. Spill and Release Control Procedures (if applicable)		
d. Work Permits & Procedures		
i. Scaffolding Permits		
ii. Confined Space Entry Permits		
iii. Cutting, Burning and Welding		
iv. Hot Work Permit		
v. Lifting and Rigging Permits		
vi. Special Equipment Operating Permits		.

145

Form 9.1 Pre-Job Planning Checklist (cont.)

	ACTION BY	COMPLETED
IV. Site Layout:		
1. Space Requirements		
a. Office Space		
b. Rest Facilities		
c. "Break" Facilities		
d. Material Laydown & Storage Space		
e. Flammable Material Storage		
f. Equipment Maintenance & Storage		
2. Traffic Control		
a. Loading/Unloading and Staging Zones Defined		
b. Traffic Flow Patterns Established and Marked		
c. Roadways, Gates or Doors Marked		
3. Postings		
a. Warning Signs per OSHA Requirements		
b. Evacuation Routes		
c. "No Smoking" Signs		
4. Custodial Service		
a. Waste Disposal		
5. Support Elements		
a. Area Lighting		
b. Noise Control & Abatement		
c. Physical Barriers to Separate Project Work from Other Operations		
d. Electrical Power		
e. Potable Water		
f. Fire Water		
g. Disposal of Waste		
V. Material, Equipment and Support: Verify the readiness of support equipment.		
1. Personal Protective Equipment		
a. Contractors Advised to Provide PPE		
b. Suppliers Identified if Needed		
2. General Use Equipment Identified		
a. Inventory Available in Stores		

Form 9.1 Pre-Job Planning Checklist (cont.)

		ACTION BY	COMPLETED
3.	Communication Mechanisms in Place		
	a. Two-Way Radios		
	b. Pagers		
	c. Telephones		
	d. Public Address System		
	e. Alarms		
4.	Fire Protection & Fire Fighting Equipment		
	a. Fire Extinguishers		
	b. Emergency Response Agreement		
	c. Fire Exits Clearly Marked		
5.	Emergency Preparedness and Support		
	a. Emergency Response Agreement		
	b. Information in Orientation		
VI.	**Education and Training**		
1.	Employee Orientation		
2.	Job-Specific Training		
VII.	**Equipment and Tool Inspection**		
1.	Heavy Equipment (test, inspection & certification)		
2.	Tool inspection program		
3.	Electrical Safety Inspection		
VIII.	**Communication & Publicity**		
1.	Communications		
2.	Reporting of Significant Occurrences		
3.	Safety Program Awareness		
4.	Other		

Form 9.2 Contractor/Guest Safety Orientation Checklist

ORIENTATION DATE : _____

INSTRUCTOR : _____

SUBJECT	ADDRESSED
1. Project/Safety Organization	
2. Company Safety Policy Statement and Copy of Rules Provided and Explained	
3. Hazard Communication	
4. Job-Specific Hazards	
5. Personal Protective Equipment Requirements (Hard Hats, Boots, Glasses, Respiratory, Hearing, Etc.)	
6. Accident Reporting (First Aid Location)	
7. Safe Work Permits (Confined Space, Hot Work, Lockout/Tagout)	
8. Vehicular Usage and Parking	
9. Eating/Drinking/Smoking	
10. Disciplinary Program	
11. Sanitation and Housekeeping	
12. Waste Disposal and Dumpster Usage	
13. Emergency Response Guidelines (Incident, Evacuation, Tornado)	
14. Fire Alarms/Safety Showers/Eye Wash Stations	

Violation of any safety rules, regulations, and procedures and/or policies will result in a warning or removal from the job-site property. Your safety is our concern, but your responsibility. Please adhere to the items listed above.

THE SUBJECT MATTER LISTED ABOVE HAS BEEN EXPLAINED TO ME AND I UNDERSTAND IT.

(**Signature** and **Social Security Number** of all attendees on reverse)

Form 9.3 Job Safety Assessment

This form is to be completed by the contractor and submitted for approval prior to work commencement. When completed, this form is to be maintained at the work site, while work is performed.

Job Title

Contract No. Building/Area

Contractor	**Facility**
Company Name: _____	Project Manager: _____
Project Manager: _____	Phone: _____
Phone: Page: _____	Division Safety Rep.: _____
Superintendent: _____	Phone: Page: _____
Phone: Page: _____	Facility Field Engineer: _____
Safety Rep.: _____	Phone: Page: _____
Phone: Page: _____	Other: _____
Designated Competent Person **For Work to be Performed**	Phone: Page: _____
	☐ Approved
_____	☐ Approved as Noted
Confined Space (29 CFR 1926.21),	
Excavation (29 CFR 1926.650),	Safety Engineering _____
Scaffolding (29 CFR 1926.451 (b))	Date: _____

- The contractor safety representative must hold an orientation with all employees prior to work, identifying the hazards to their scope of work, and have each person sign the attached signature sheet.

- Identify location of telephones and designated tornado shelters in relationship to the work site and provide the following telephone numbers: telephones, dial 9-1-1; cell telephones, dial (630) 252-1911.

- Emphasize compliance with OSHA 29 CFR 1926.

- Utilizing the format on the following pages, identify hazards and safety precautions/procedures to mitigate hazards.

JOB SAFETY ANALYSIS	JOB TITLE PAGE OF JSA NO.		DATE: ☐ NEW ☐ REVISED
ORGANIZATION:	**ANALYSIS BY:**	**REVIEWED BY:**	**APPROVED BY:**

Scope (Description) of Work

Basic Safety Rule Reminders

- Inspect all tools and equipment for OSHA compliance before use.
- Fall protection is required when working at heights (standing surface) above 6 feet and a handrail or other fall protection is not provided.
- Flag work areas and post warning signs.
- Ground fault circuit interrupters (GFCIs) are required on all 110- and 120-Volt receptacles.
- Stairways, passageways, and accessways must be kept free of materials and equipment and orderly housekeeping shall be maintained.
- Report all injuries, illnesses, and near misses to the Facility Field Engineer.

REQUIRED AND/OR RECOMMENDED PERSONAL PROTECTIVE EQUIPMENT

Phase of Work/Basic Job Steps	Safety Concerns/Potential Hazards	Recommended Action or Safety Procedures

Continued on next page

| JOB SAFETY ANALYSIS | JOB TITLE | DATE: | ☐ NEW |
| | PAGE OF   JSA NO. | | ☐ REVISED |

REQUIRED AND/OR RECOMMENDED PERSONAL PROTECTIVE EQUIPMENT

Phase of Work/Basic Job Steps	Safety Concerns/Potential Hazards	Recommended Action or Safety Procedures

Material Safety Data Sheets (MSDS)

Hazardous materials used on this site   Location(s) of MSDS:
are/will be

Hazardous materials	Location(s) of MSDS
1.	1.
2.	2.
3.	3.
4.	4.
5.	5.

Review of Emergency Routes and Assembly Point(s) (Basic Information)
(Use separate sheets as necessary.)

_____

_____

_____

_____

Signature Sheet
Job-Specific Safety Orientation

Contractor: _____   Building/Area:_____

Contract Number: _____   Job Title: _____

Contractor's

Superintendent: _____   Safety Rep.:_____

"Safety information relative to this job/project has been reviewed with me by my
company's Safety representative."

Name (please print)	Badge No.	Signature	Date

Chapter 10

Role of the Field Engineer

I.   Introduction

It is the field engineer's job to make sure the subcontractor clearly understands the technical requirements of the work and to monitor the contractor's integration of safety into their work execution. The field engineer is the contractor's "Go-To-Guy" for information regarding where to go and when to be there to perform work. Without this support the contractor's efforts will not be applied efficiently during the often-short period they are on the job site. In some environments without the guidance of a facility representative contractors can unknowingly place themselves in potentially dangerous situations.

Even with the pre-screening process, site-specific safety program reviews, and safety orientations, the host employer cannot be certain that the contractor will perform within defined job-site safety standards. At this point the human element must be introduced, this is the field engineer. Contrary to the fact that many contractors believe or would like to believe that their host employer is solely responsible for all work-related safety and health matters, it is each contractor's (employer) responsibility to see to the safety of his respective employees. The role of the field engineer is to ensure that the contractors understand the job-site hazards and are clear as to what is expected of them and to coordinate intercontractor efforts and finally to oversee their compliance with the contract-specified safety requirements.

II. Role of the Field Engineer

The role of the field engineer can be boiled down to the following functions:

- Identifying site-specific safety hazards to contractors,
- Establishing that the contractors recognize the hazards and are prepared to deal with them,
- Coordinating the interfaces between contractors,
- Coordinating the interfaces between contractors and operating facilities, and
- Verifying that the contractor is performing to agreed upon contract requirements.

To illustrate the importance of the role of the field engineer, let's take a look at an actual work project. A general contractor remodeling a laboratory subcontracted the fabrication and installation of a roof-mounted fan unit for a fume hood. The facility's owner was very much aware of the potential hazards associated with mobile crane operations in the vicinity of operating facilities and insisted that all lifts be closely monitored.

At 7:30 a.m. on a Thursday morning the field engineer was waiting at the designated lift point for the arrival of the vendor's fan unit and a rental crane. Considerable coordination and planning had gone into preparing for this operation. Although this was not a critical lift, detailed directions had been provided to the contractor to ensure he understood the site-specific requirements to complete the lift. The field engineer was hoping the lift would be done before the daily deliveries started to arrive at the loading dock adjacent to the operation. Had the contractor understood and acted on all the discussions associated with the work?

The following details were addressed in the planning meeting with the subcontractor providing the fan unit:

- The rental agency was to provide a certification of inspection for the crane provided and the operator was to have a current medical and operating certificate issued by the National Commission for the Certification of Crane Operators.
- The required boom length calculated by the contractor had been confirmed.
- The fan unit weight was determined.

- The radius of operation, required boom length, the type of block and rigging needed, the weight of the load, and the required boom attachments had all been evaluated to establish the size of the required crane.

- The lift site had been evaluated to ensure that underground piping, sewer lines, tanks and utility service-ways would be avoided. Matting requirements for the outriggers were defined to avoid damaging the parking lot paving.

- The optimal crane position had been identified and barricaded. Considerations of the crane's location included ensuring that the load could be brought right up to the crane and that the load would not have to be lifted over critical equipment.

- A lift plan had been developed to detail how the lift would be made and facility structures avoided.

The field engineer was present to observe the crane inspection and to make sure the lift was conducted according to the agreed upon lift plan. At 7:40 a.m. both the crane and fan unit arrived. A quick inspection of the crane revealed that it did not have the jib needed to reach the roof location for the fan unit.

The crane was sent back and another one arrived at 9:00 a.m. This one had the correct jib. However, the operator's medical certificate had expired and he did not have an operating certificate as specified. The crane rental firm brought in another operator who did have the proper certification. He arrived at 10:30. By this time seven people had been standing around for three hours waiting for the lift to start.

The lift plan was reviewed with the new operator who then set up the crane. The slings were secured to the fan unit then the owner's safety representative was called in to conduct the inspection of the crane and lift configuration. On completion of his review he asked if the lifting lugs that had been welded onto the unit by the contractor would support the load of the unit. What evidence did they have to substantiate their assertion that it would? Now what should the contractor do? The field engineer offered the contractor a couple of suggestions to help the contractor demonstrate that the lugs would support the lift. The job was completed at 11:30 a.m. without a safety mishap.

In this example we see the value of the field engineer in the contractor's work process. What further delays would have been experienced if the contractor had not planned the work? Apparently the contractor failed to clearly communicate the details of the lift and the facility owner's requirements with the rental agency, in spite of the specificity with which the field engineer had discussed the details of the lift with the contractor.

To effectively perform the field engineer function, the individual designated to fill this role should be conversant with requisite safety regulations and have a practical working knowledge of safe work practices. This will enable the individual to identify practical solutions to safety issues and mitigation of safety hazards.

III.   Work Release Meeting

Before approving the contractor's "Notice to Proceed," the field engineer must be confident that all required planning has been completed and requisite documentation is in place. This applies to the start of a $5 million process plant expansion or the installation of a fan unit on the roof of a building.

The field engineer responsible for overseeing the work should hold a meeting with the contractor's supervisors as soon as the following actions have been completed but before the start of any contracted work:

- Contract is signed;
- Performance, payment bonds and insurance have been secured;
- Safety and Health program approved;
- Job-specific safety analysis reviewed; and
- Schedule is approved.

At this meeting the field engineer should confirm that the key contractor and facility personnel have a clear understanding of the contract's scope, as well as job-specific hazards and safety requirements. The meeting agenda should review how the contractor and facility will do business with each other, the procedures the contractor will follow, and any facility-specific requirements that must be met before the start of and during the performance of the work. A meeting agenda such as in Figure 10.1 should be distributed to the participants before the meeting.

Meeting minutes with a summary of items discussed should be distributed to all attendees along with a list of principal contacts for the project (for both contractor and facility) with telephone numbers for work day and off-hour emergency contacts. The minutes should be transmitted within five (5) working days following the meeting.

IV.   Equipment Inspections

All heavy equipment brought on site should be inspected to determine that it will operate throughout its expected stay. The field engineer responsible for the inspection should be able to call on an individual with the technical expertise to conduct the inspection if he is not familiar with a specific piece of equipment. The purpose of this inspection is not to certify the equipment, but to verify that it has been certified and appears to be in good working condition.

For example, an equipment inspection might include verification of

- Seatbelts
- Lights
- Horns
- Back-up alarm

Pre-Work Release Meeting Agenda	
Introduction	Project Manager (PM)
Scope of work	PM
Status of required documentation	
• Acceptance of contract documents	
• Site-specific safety program approved	
• Job Safety Analysis has been reviewed	PM
Job-specific safety & health requirements	Safety Coordinator
• Inspection of equipment brought to site	
• Material Safety Data Sheets submitted	
• Emergency phone numbers provided	
• Attendance at site-specific orientation	
• Designation of safety representative with at least 10-hour OSHA Training has been received in writing.	
• Bulletin board with all pertinent information clearly posted.	
Security requirements	Security
Permits and hold points	Field Engineer
Contractor reporting requirements	Field Engineer
• Inspection and acceptance requirements	
Work laydown area assignments	Field Engineer
Utilities and temporary services	Field Engineer
• Confirmation that permits have been secured, signed and will be posted	
Review coordination of contractor's work	Field Engineer
• Confirmation of project contact points	
• Coordination meeting schedules	
• Reminder of when reports are due	
• Weekly Manhour and Injury Report	
Change and extra work procedures	PM
Progress payments	PM
Questions	Contractor
Summary	PM

Figure 10.1 Pre-work release meeting agenda—example.

- Windshield in place and free of defects
- Absence of leaks
- Fire extinguisher
- Roll-over protection (where necessary)
- Cranes should have:
  - Anti-two-blocking
  - Cables in good condition
  - Inspection certification
  - Operator medical exam
  - Operator certification
  - Load chart
  - Angle indicator
  - Motion alarm
  - Proximity alarm

Invariably, very sorry-looking equipment shows up on new projects. However, once a few cranes and heavy pieces of equipment have been denied access to the job site based on the justified determination that the equipment does not meet job-site standards, the word will get around and serviceable equipment will arrive at the gate. That other equipment will be sent to less demanding clients. The last thing a job-site manager wants on site is a piece of equipment that may fail in the midst of a work activity, disrupting the workflow of the project. It is better to invest time in work preparation and ensuring that everything is in place than to have to take remedial steps to correct a failure.

V.   Tool Inspections

Contractors are usually expected to provide their own tools and equipment. It is expected that these should be in good working order. After all, it is each employer's responsibility to ensure that the tools provided or used by his employees are in good working order. Sadly, many craftsmen employ tools that are in poor condition. Contractors will argue that the condition of their workers' tools is their responsibility and not that of the host employer. Contractors will suggest that inspecting their tools is akin to directing the means and methods of their work. The fact is that high levels of performance can only be achieved if the craftsmen have the requisite tools and that they are in good working order. Although it is up to the contractor to deliver their work on time, it is in the job-site manager's interest to avoid all foreseeable work interruptions. The tool inspections conducted by the field engineer are one more means of determining the level of safety practiced by the contractor. The host employer is not absolving the contractor of any responsibility by inspecting the tools. If, by chance, a defective tool is observed, the field engineer can point it out to the contractor. Depending on the number of substandard pieces of equipment identified, the field engineer will be able to develop a feel for the level of oversight that will be required of the contractor.

By this point I hope the reader has reached the conclusion that implementation of an effective safety program is not achieved by meeting the letter of 29 CFR 1926 or 1910. An effective safety program, the avoidance of injuries and fatalities, is achieved by identifying and eliminating hazards. Putting the requirement on the contractor through the contract will not work without active involvement in the safety program. It is in the interests of both the host employer and the contractor to realize that a contractor injury resulting from a defective tool will be just as disruptive as delay of material arrival or any other work-related interruption. Those individuals concerned about litigation will not have to deal with the issue if there are no injuries.

The fact is that craftsmen are often pleased that someone will intercede on their behalf to get substandard tools and equipment replaced. However, in the interest of self-preservation workers may not raise the issue regarding tools that are not in good working order if it means keeping their jobs.

The reason contractor tools may be in poor condition may also be the fact that the craftsmen don't recognize the hazard or have grown to accept the condition of their tools. The following is a list of the more frequent equipment deficiencies identified during my tool inspections:

- Ladders without safety feet and damaged rungs
- Handles missing on files
- Nicked and frayed extension cords
- Ground prongs removed on extension cords and electrical tools
- Cracked housings on power tools
- Mushroomed heads on chisels
- Cracked handles on hammers
- Clips missing on compressor airline hoses

Defective tools and equipment should be tagged out-of-service immediately and removed from the job site by the contractor by the end of the shift. The inspection of tools should be performed following the workers' job-site orientation where they have been given notice that this process will take place. Observations should be done in a positive manner, explaining why a tool is judged to be unsatisfactory where this is the case.

VI.   Job-Site Monitoring

The degree and quality of coordination provided by the field engineer have a direct effect on contractor safety performance. The greater the support provided to the contractors the better they will understand the job-site requirements and the more productive they will be. The degree to which the field engineer emphasizes safety will directly affect the emphasis the contractor places on safety performance. Field engineers must emphasize safety in their daily communications with contractor supervisors and when they have the opportunity to interface with the workers.

VII. Inspections

The field engineer should maintain a field notebook to record daily observations. In addition to quality, schedule, performance observations, and records of job-site conditions, the field engineer should monitor safety compliance. Job-site hazards that cannot be corrected immediately should be noted. The corrective action taken to address safety issues should be recorded.

December 8, 1998

*7:30 Monitored Daily Pre-job Meeting – Superintendent reviewed yesterday's activities. He introduced a new work crew member – Sam Baker; Bill Simmons will be his buddy. The **Toolbox Talk** consisted of a review of the lifting operation planned for today to place roof air-conditioning unit.*

8:00 Pegg called and reminded me about soccer tonight. Need to buy oranges for the kids. (It's perfectly okay if some personal errands are included in the notebook if this is where the engineer makes notes to himself.)

8:30 Reminded Sam Baker (Electrical Svc. Inc) to wear his hard hat.

8:45 Conducted crane inspection for lift. Driver (Bill Williams) had his certification and medical current this time.

9:15 Handrail on 2ⁿᵈ floor by Column 198 needs to be tightened. Called John Jones (Foreman) to fix handrail.

*9:30 Told Bill Downey (Don's Piping, Inc.) to put on his safety glasses a second time. **B. Downey***

10:00 Project coordination meeting – Reviewed this morning's field observations.

Figure 10.2 Sample notebook page

Deficiencies such as masons overloading their scaffolding or missing toe boards that should be in place should be brought to the contract supervisor's attention for correction. This observation should be entered in the field engineer's notebook as a record of the conversation. It puts the contractor on notice and provides the field engineer with a record of the event. If the situation surfaces again, the field engineer has the documented basis for beginning formal corrective action.

Since in a contracted work environment it is not generally good form to give work direction to subcontractor employees, it is a good idea to include the contractor supervisor in routine job-site inspections so that observed deficiencies can be identified concurrently and corrected on the spot.

Job-site progress photos and videos provide a good visual record of job-site conditions. Photographs are indisputable evidence in the event of an argument or can be used at job progress meetings to clearly point out housekeeping, worksite conditions or specific work practices needing to be addressed. The availability of electronic cameras and computers enables photographs and overhead viewgraphs to be generated with little difficulty.

VIII.   Enforcement

Adherence to established safety standards is a significant factor in the level of safety in the work environment. An important role of the field engineer is to monitor general work practices and bring safety discrepancies to the attention of the respective employers and their employees.

Observations of undesired safety behavior should be noted in the field engineer's notebook. This is not the written notice issued the second time a substandard behavior is noted. The field engineer's record of the first approach will diffuse any argument about whether the worker was previously corrected for a similar infraction.

"Bill, this is the second time I have seen you not wearing your safety glasses this week; we spoke about it before. You know it's a project requirement to wear safety glasses. Look, I don't want to make a federal case out of this, but I am going to make a note that I am giving you a formal notice. Would you please initial in my book that we talked about this? Bill, please wear your safety glasses from now on." Document your safety conversations.

The field engineer should not feel he is being a "Nice Guy" by walking away from a safety issue that could hurt a worker later. Remember, what you condone today you will have to accept tomorrow. The objective of a proactive safety program is not to threaten workers or their jobs, but to encourage them to work safely so they will be there the next day.

IX.   Accident Investigation and Reporting

Accident investigations are generally relegated to the contractor. However, the field engineer should participate in the investigation process, even if it is just in the role of an observer. This will provide insight into why the incident occurred so as to be able to prevent a similar incident from happening again and assure that proper documentation is collected for the safety file. The investigation of occurrences (accidents) is the only way the source of the incident can be determined. Investigations should not be a hunt for the guilty perpetrator, but a determination of why the incident occurred in the first place.

X.   Progress Meetings

Ask ten people what they think of meetings and you will get ten different opinions. What they will all agree is that most meetings are poorly organized and run. Planning and coordination are keys to the success of meetings.

Produce an agenda for each meeting. This will provide focus and will demonstrate that you have given the subject some thought and that you have prepared for the meeting. An agenda will also help you maintain the focus of the meeting.

Job Progress Meeting
Agenda

Introduction
Purpose of Meeting
Monthly Accident Report
Job-site Inspection Report Results
Corrective Action Taken/Required
Safety Coordination
Next Week's Activities
New Business
Next Meeting Date/Time
Adjourn

Figure 10.3 Job progress meeting agenda—example.

Schedule the meeting for a specific duration and let the attendees know what that is. Then stick to your timetable. If you don't get everything addressed in the allocated period of time, identify what those issues are and propose to address them at your next meeting. Don't go over your meeting time unless it is absolutely essential and then only with the permission of the participants.

I suggest that meetings never last more than one hour. Your participants may not come back again if the meetings are longer than that or you consistently run over your scheduled time. Remember, they may have something else scheduled following your meeting. If you can address your subject in less than the time you have scheduled, state that you have done so and then adjourn. The attendees will appreciate this and will be more willing to attend the next time you schedule a meeting.

Post the objective of the meeting in a prominent location in the room. It helps everyone maintain his or her focus on the meeting's objective. If the conversation strays, the objective helps draw you back to the focus.

XI.   Job Safety File

The field engineer should maintain a job safety file. This should contain the records that will be transferred to the client or originator when the work is done. It should contain all the information the facility operator needs to know regarding the construction of the facility.

- As-built specifications and drawings,
- Design criteria,
- General details of construction methods and material used,
- Potential hazards such as locations of post-tensioned tendons,
- Manuals produced by equipment suppliers and specialist contractors which outline operating procedures and maintenance schedules for installed equipment,
- Requirements for cleaning and repair of components,
- Locations of utilities and service connections, and
- Sources of potential risk to operations and maintenance personnel identified during the contracted work process should be brought to the client's attention and highlighted in the file, e.g., confined spaces.

The field engineer should discuss the proposed contents of this file with the originator/client before the work begins to identify if there is any additional information that he would like included in the file. The field engineer should collect the project/job information as the work evolves.

XII.   Summary

Prudent host employers recognize that it is not enough to simply exercise reasonable care in the initial selection and employment of specialty contractors to perform contracted work. In order to effectively prevent accidents when working with contractors, the host employer undertaking the project must monitor the work for the duration of the contract. The individual generally assigned the responsibility to monitor the contractors is the field engineer.

Chapter 11

Procedures and Permits

I.   Introduction

Procedures are an important element in an effective safety program. They are the means by which to document an accepted step-by-step approach to perform high-risk work safely. Procedures serve to identify known hazards associated with specific work activities and to define the process by which to control those hazards within the context of the work being performed. When developed with the benefitof the collective experience of seasoned tradesmen and technical specialists, procedures offer both employers and their workers the assurance that the work can be completed safely and in compliance with policies and regulatory requirements.

Permits control high-risk work activities encompassed by procedures and provide a structure to ensure that requisite training and controls have been implemented prior to the authorization being given to start work.

II.   Procedures

As less than 20% of construction contractors have had the benefit of learning their trade through a formal program, the question that begs to be answered is where did the rest of them develop their trade skills. Generally, it is on-the-job training. Bill learned from Dave, who learned from Robert. But, do they really know how to perform the work safely. Procedures provide a framework to guide supervisors and workers when planning to perform potentially high-risk tasks such as entry into a confined space. Procedures identify who is expected to do what, how the work should be performed, what controls are to be in place, what the role of the attendants is, and who is expected to provide emergency response support.

Procedures prepared for a broad spectrum of activities are commonly referred to as Standard Operating Procedures. Of particular interest here are those procedures that address potentially hazardous work such as critical lifts, working at heights, working in confined spaces, electrical hot work, etc.

A.   Standard Framework for Development of Procedures

TITLE: Subject being addressed

1. AUTHORIZATION

Review Period – The effective life of the procedure.

Developed, Reviewed, and Approved – Identify the individuals who developed, reviewed, and approved the procedure and on what date. This provides a link back to the author should clarification be required. The dates provide an indication of how current the procedure might be and how close the need for a revision of the procedure might be.

2. REFERENCES: The reference authority upon which the document was based in the event the reader would like more information.

3. INTRODUCTION

3.1. Purpose   Procedures provide consistent application of safe work practices among workers. Within this section define the reason for the procedure, such as providing specific direction for the operation of a powder-actuated tool or entry to a confined space.

3.2. Scope   Procedures should be developed for all high-risk operations, tests, and abnormal or emergency situations. They should provide a description of what the procedure encompasses and define what is not included.

3.3. Applicability   This section defines who is expected to follow the procedure and who is required to support it.

3.4. Definitions   Define all terms used in the procedures to ensure a common understanding of them.

3.5. Program Components

A. Warning notices
B. Work steps
C. Identified hazards
D. Hazard control measures (including safe work procedures)
E. Operating and rescue procedures
F. Training and education requirements

The extent of detail in a procedure will depend on the complexity of the task, the experience and training of the workers, the frequency of performance, and the significance of the consequences of error. Although a complete description of a system or process may not be needed, procedures should be sufficiently detailed so that the required functions can be performed without direct supervision.

B. Approvals

Supervisors and management should be engaged in the review and approval process so that they clearly understand the requirements of the procedure, the actions prescribed in the procedures, and the safety controls needed for the personnel executing the procedure.

The author's supervisor should approve the technical content of each procedure. The supervisor's signature on the procedure certifies that the procedure correctly describes the work to be performed, and the procedure can be safely carried out as described. It approves of the clarity of the procedure, and confirms that the procedure can and should be performed by workers as outlined.

- Procedures should be easily understood, and actions should be clearly stated.
- Procedures should contain only one action per step.
- Procedures should contain sufficient but not excessive detail.
- Warnings, notes, and cautions should precede the step to which they apply, they should be easily identifiable, and should not contain action statements. The probability of missing an action step increases when it is included in a warning, note, or caution.

Procedures must be technically and administratively accurate (i.e., the instructions and information should be correct, referenced documents should be correctly identified, and necessary instructions present to guide the user when executing the procedures). The sequence of procedural steps should conform to the normal or expected operational sequence of activities.

C. Review of Procedures

Periodic changes and revisions to procedures are necessary to ensure that they reflect current practices and requirements. The review and approval process for each procedure change or revision should be defined when the procedure is developed.

- Applicable procedures should be reviewed after an unusual incident (such as an accident, significant worker error, or equipment malfunction).

- Appropriate procedure changes and revisions should be initiated when procedure inadequacies or errors are noted.

- Important information regarding changes or revised procedures should be communicated to appropriate workers prior to the next time that activity is performed.

- The frequency of reviews will vary with the type and complexity of the activity involved. A proposed revision schedule should be established when each procedure is developed.

- During reviews, procedures should be compared to source documents to verify their accuracy.

D. Documentation

Define the documentation requisite to record and track training, monitoring measurements, observations, etc. associated with the procedure and its execution.

E. Training

Establish a worker training program that addresses the following:

- Contents of the procedure
- Nature of hazards present
- Signs, symptoms, and consequences of exposure if applicable
- Safe operating procedures
- Emergency communication
- Immediately terminate work when alarm, warning sign, or other uncontrolled hazard begins

III. Permits

Permits provide the structure to ensure that all hazards have been controlled, that workers understand the work to be performed, that approvers[1] are alerted to de-energize facility controlled sources of energy and that emergency responders

---

[1] Approvers – The individuals who are responsible for issuing dig permits, de-energizing power to lines and other facility power sources.

are aware of the potential need for response before high-risk activities are permitted to start. Permits define who is responsible for controlling the work, under what circumstances work is to be aborted, confirmation of agreed-to safety precautions and emergency response reactions required.

Permits also give the field engineer notice of the need to coordinate with other contractors and individuals working nearby about any high-risk activities that may be about to begin. Once these conditions have been satisfied, the permit provides documentation of the authorization to start the work.

A. Permits should include at least the following information:

1. Location of work to be performed
2. Date
3. Requested start time
4. Time work is to end
5. Time of issue of permit
6. Names of workers assigned to perform work
7. The name and position of the person authorizing or in charge of the activity
8. Description of hazards known or reasonably expected to be present where the work is to be performed
9. Specification of required means of energy isolation, cleaning, purging or inserting to be done before entry to remove or control hazards
10. Testing or monitoring results of conditions before the work begins (e.g., zero energy in lockout/tagout or air quality prior to a confined space entry)
11. The individual who performed the tests
12. Personal protective equipment required:
    a. respiratory protection
    b. clothing
    c. etc.
13. Identification of special work practices or procedures to be followed
14. Planned emergency response and rescue retrieval method, where appropriate
15. Description of hazards or conditions that may reasonably be expected to be generated by the permitted work process, e.g., noxious fumes
16. Communication system (e.g., two-way radio, telephone for 911)

B. General Procedure

Permit forms must be filled out completely, all required signatures must be in place and requisite attachments, such documentation of sampling, must be collected before work is permitted to begin. In the interest of expediency it is sometimes the case that requesters will assume that one or another of the authorized approvers will sign off on the permit and decide to start work without all signatures. Although it might well be the case that none of the approvers would have a problem signing the permit, the absence of their specific endorsement on the permit might result in some crucial safety precaution being omitted.

Deviations from permitted approvals present the potential for safety complications. For example, if the facility's safety representative is not notified of an early start of a proposed open flame operation, and smoke sensitive detection systems are not deactivated in time, the resulting alarm may well direct the evacuation of the entire operating facility, resulting in a very unhappy facility owner. What if a hydrocarbon line-break in progress nearby has not been completed when the open flame operation starts; this could result in serious complications.

Because conditions are continually changing in contracted work environments, permits should only be valid for a limited period of time. When a permit has expired and an extension is needed, the work conditions should all be evaluated again.

C. Permit Signatures

The review and sign-off on the work to be performed under the permit requirement are important elements in the safety permit process.

Signatures of individuals selected to perform the work on the permit should indicate that they understand how the work is to be performed and the permit requirements.

Signatures by supervisory personnel should indicate that they have reviewed the permit procedures with their workers and are satisfied that the workers understand how the work is to be performed.

In complex situations and situations where failure to follow safe work practices or to control work environment conditions could lead to serious injuries, illness, or death, such permits may require independent evaluations by more than one person, in which case each of those people should sign the form.

Signatures by specialists such as Industrial Hygiene, Fire Department, and other properly trained individuals indicate that they have evaluated the anticipated work activity, made appropriate measurements or acknowledge that energy sources have been disabled.

Permit systems can be easily defeated. Individuals can sign the permits without fulfilling their responsibility to inspect or perform measurements prerequisite for the permit. Each individual signing off on the permit must be held accountable for having exercised his respective judgement or measurements before the activity is allowed to proceed.

D. Supervisor Responsibility

The supervisor of the work to be performed is responsible for

- Determining when the planning, procedures, and permits are required;
- Initiating and preparing permits;
- Ensuring that no work begins until the requirements of applicable permits are fulfilled;

- Ensuring that work conducted under his/her supervision is performed safely and in compliance with issued permits; and

- Coordinating with others who may be affected by the work to be performed.

The completed safe work permit forms should be posted in the job-site office and at the specific site where work is to be performed. When the permitted work is completed, the work site should be clear and free of safety hazards. Where this is not the case, the permit applicant should be required to restore the area to a safe condition.

IV.   Types of Permits

The following is a list of safe work permits typically in place in operating facilities. This should not be construed as being all-inclusive. Each facility will have its own permit requirements and implementation procedures.

- Confined Space Entry
- Digging
- Fire Protection System Impairment
- Movable Structure Siting
- Critical Lifts
- Movement of Heavy Equipment
- Radiation Sources Work
- Lockout/Tagout
- Electrical Hot Work
- Work Entry Clearance
- Hot Work/Open Flame and Spark-Producing Operations (Example Form 11.1)

A. Confined Space Entry Permit

A confined space entry permit is required any time work is to be done in a confined space. A confined space is an enclosed area that has the following characteristics:

- Has limited or restricted means for entry or exit (for example, tanks, vessels, silos, storage bins, hoppers, vaults, and pits are spaces that may have limited means of entry);

- Is large enough and so configured that an employee can bodily enter and perform assigned work, but is not designed for continuous employee occupancy or is not routinely occupied;

- Contains or has potential to contain a hazardous atmosphere;

- Contains a material that has the potential for engulfment;
- Has an internal configuration of inwardly converging walls or sloping floor; or
- Contains any other recognized serious safety or health hazard.

Work in confined spaces must not proceed until all involved personnel have followed the procedures and complied with the facility-specific requirements. The host employer or job-site manager must apprise the contractor of the precautions or procedures, if any, that the host employer has implemented for the protection of employees in or near permit spaces where contractor personnel will be working, before work begins.

B. Digging Permit

Digging may not be a great concern at the start of green-field projects; however, permitting becomes increasingly important as projects evolve and underground utilities are installed. As-built underground utility drawings should be developed as projects evolve and updated with each digging permit requested. Project personnel often rely on their collective memory of where utilities were buried, but as projects draw to an end and project personnel move on, the recreation of this information may be costly if clear records are not maintained.

Safety considerations associated with the control of digging permits are

- Assurance that no buried utilities will be unwittingly disturbed.
- Coordination of the disturbed area and control of traffic in the affected area.
- Assurance that a competent person has been identified to monitor the excavation related activities.

C. Fire Protection System Impairment Permit

Fire Protection approval must be obtained prior to the planned impairment of any fire protection system or fire protection system component(s) in existing or recently commissioned facilities. False alarms are an aggravation, but more serious is the loss of control of the system's status if multiple organizations are activating and deactivating fire protection systems.

D. Movable Structure Siting Permit

The facility owner may have a site-specific requirement that trailers and other movable structures be located in accordance with established procedures and local building codes. In the midwest where high wind conditions are common, one host employer insists that trailers are strapped down. A permit may be required for each movable structure, or when grouped, for each group of movable structures.

E. Heavy/Critical Lifts

Critical lifts present a very real threat to project costs and schedules if not executed properly. Detailed planning for all heavy lifts should begin during the planning stage of the project. (A heavy lift is defined as any lift exceeding 80% of the capacity of the lifting equipment, or requiring the use of two or more pieces of equipment in unison to make the lift.) Similar planning should be done for any lifts made over the owner's facilities and equipment.

The following are some lift planning considerations, but much like everything else in construction the details depend on the specifics of the lift, its location, the equipment to be used, site constraints, etc.

- Equipment to be used and their load capacity

- Position of equipment during lift

- Weight and center of gravity of load

- Location and capacity of lifting lugs on equipment to be lifted

- Boom length, angles and radius of lift

- Bracing and reinforcement required on equipment to be lifted

F. Movement of Heavy Equipment

Sites with low overhead process lines or weight-sensitive underground utilities may have permit requirements for the movement of heavy equipment.

G. Electrical Hot Work Permit

Work involving potentially hazardous electrical circuits must not be initiated unless all involved personnel have followed defined procedures and complied with the lockout/tagout requirements defined by the site host and 29 CFR 1926 for working on energized electrical circuits (working hot electrically). Every effort should be made to de-energize and lockout/tagout electrical equipment before considering any requests to work hot.

H. Work Entry Permit

Established operating facilities often have Work Entry Permit requirements. This permit alerts the facility manager to work outside of routine operations. Work entry permits are often required for maintenance, service contracts, and any other work being done at the facility that is not part of routine operations. Each of these activities has the potential to disrupt facility operations if proper coordination of activities fails. The work entry permit is a means of informing the facility manager of the potential generation of noxious fumes or dust. The facility manager is then in a position to define control measures such as fans, or alternate work hours to minimize negative effects or disruption of facility operations.

I.   Hot Work/Open Flame Permit

This permit establishes fire prevention and safety requirements for open flame or spark-producing operations such as welding, cutting, brazing, flame soldering, and melting in which open flame, arc, and sparks present the potential for fire, explosion, and burns.

V.   Summary

The presence of procedures and permits to control known hazards may initially be viewed as being more of a bureaucratic exercise than one of promoting a safe workplace. However, the process of developing procedures is very instructive, and one that forces the developers of the procedures to evaluate each work step. Similarly, permits require that their applicants give thought to the work to be done, rather than leaping into the work without planning and preparing for it.

APPENDIX 11.1

Procedures

It is in the interests of the job-site manager to provide specific guidelines for activities that present probable risk of injury. Expectations and requirements that are locked in the heads of the job-site safety representative are of little value to contractors who then feel that they are required to play a guessing game with the safety representative regarding what is expected of them. When procedures and requirements are documented, there is a clear basis for discussion to resolve potential conflicts. The following is an example of a Safety Procedure that clearly defines requirements and expectations.

Example Procedure
Lockout/Tagout Procedure

1. AUTHORIZATION

Review Period:	1 year	
Developed by:	Jim Lang	11/5/98
Reviewed by:	Rick Janik	11/30/98
Approved by:	Jim Bennett	12/10/98

Provider: Environment, Safety & Health Office
Related ES&H Policy: Policy 6-4, Worker Safety and Health
Document #: WF-16

The Lockout/Tagout Program provides specific details for compliance with safety requirements established by Policy 6-4, Worker Safety and Health. All responsibilities, general rules, and procedures from Policy 6-4 apply to this program.

2. REFERENCES

29 CFR 1910.147 (OSHA) — "The Control of Hazardous Energy (Lockout/ Tagout)"

29 CFR 1910.331 thru .335 (OSHA) — "Electrical Safety-Related Work Practices"

Woodlake Electrical Safety Program — Doc.# ESH-20

Policy 6-4, Worker Safety and Health.

3. INTRODUCTION

Table of Contents

1. Purpose
2. Scope
3. Applicability

3.1 Purpose

The Lockout/Tagout Program and example procedures have been developed to ensure that all equipment is deenergized and physically removed and isolated from external and internal power sources prior to operations, such as maintenance or construction activities, in which accidental startup or the release of energy could cause injury. The primary purpose of the program is to protect personnel from injury. A secondary purpose is to protect equipment from damage. This program was developed to meet the operational needs of Woodlake Refining while being in full compliance with applicable regulatory documents referenced in Section 2 above.

3.2 Scope

The Lockout/Tagout Program applies to all energy sources including, but not limited to, electrical, steam, hydraulic, mechanical, chemical, thermal, and pneumatic. Internal energy sources such as charged capacitors, batteries, wound springs, raised loads, thermal sources, etc. are also covered under this program and must be properly relieved or restrained before activities begin.

This program encompasses facility-wide activities and applies to all permanent and temporary employees. Contractors working at Woodlake facilities must also adhere to this program or be able to demonstrate that they have a program in place that provides an equivalent level of protection.

3.3 Applicability

The Lockout/Tagout procedure applies to all personnel performing work at the Woodlake Refining facility. This applies to direct hire and contracted personnel. In accordance with the requirements of 29 CFR 1910.147 paragraph (f) any contractor required to control an energy source at this facility will discuss the details of the lockout with his designated field engineer.

3.4 Definitions

Administrative Tagout: The placement of "Out of Service" or "Authorized Personnel Only" control tags to inform personnel that operating restrictions have been placed on a piece of equipment. Administrative Tagout can be applied without Administrative Lockout. Administrative Tagout shall never be used for isolation of an energy source.

Administrative Lockout: Locks that are applied to equipment control circuits or energy-isolating devices for the purpose of preventing operation by unauthorized personnel. A control tag which states "Caution, Operation of this Equipment by Authorized Personnel Only" or "Out of Service" must accompany padlocks used for administrative purposes. Administrative Lockout shall never be used for isolation of an energy source.

Control Circuit: A circuit that contains switching devices that control the activation and/or specific operations of a piece of equipment. Control circuits are poor locations to apply lockout devices since they do not necessarily disconnect input power.

Control Tags: Tags used to identify the operational status of a piece of equipment, system, etc. Three types of preprinted tags are authorized for general use at Woodlake. All have spaces for specific information to be added, such as name, phone number, date, etc.

"DANGER DO NOT OPERATE": These control tags are used to identify equipment or controls that are subject to a Lockout/Tagout and MUST NOT be operated. They are colored red, black, and white.

"OUT OF SERVICE": These control tags are used to identify equipment that is removed from service for an administrative reason but is not subject to Lockout/Tagout. They are colored yellow and black.

"AUTHORIZED PERSONNEL ONLY": These control tags are used to identify equipment that has restrictions in effect concerning who is authorized to operate it. They are colored yellow, black, and white.

Disconnect Switch: A lockable electrical switch that physically disconnects equipment from its input electrical power source.

Energy-Isolating Device: A mechanical device that prevents the transmission or release of energy. Examples include a circuit breaker, a disconnect switch, a flow control valve, a slide gate, a blind flange, a block, and any similar device used to block or isolate energy. The term does not usually include push button or other control circuit-type devices.

External Energy Source: Energy sources which are external to equipment such as electrical, hydraulic, pneumatic, gas, vacuum, high temperature, cryogenic temperature, mechanical, etc. that could cause harm to personnel or equipment.

Internal Energy Source: Energy sources such as capacitors, accumulators, air surge tanks, batteries, hydraulic line pressures, wound springs, etc. that are internal and could potentially be released and cause injury after all external energy sources have been disconnected and secured.

Lockout: The placement of a locking device, such as a padlock, on an energy-isolating device, thereby preventing the energy-isolating device and the equipment being controlled from being operated until the lockout device is removed. Lockout must always be accompanied by tagout.

Tagout: The placement of "Danger Do Not Operate" tags on an energy-isolating device to inform personnel that the energy-isolating device and the equipment being controlled MUST NOT be operated. Tagout must always accompany a lockout, and may only be used alone if the energy-isolating device cannot physically be locked out and when alternative isolating procedures are used (i.e., removal of a circuit breaker or fuse).

Zero-Energy State: A term that applies to equipment or systems status in which all hazardous energy sources have been disconnected and secured and all internal energy sources have been relieved or restrained in a safe manner.

3.5. Program Components and Assignments

A.   General Requirements

This program encompasses facility-wide activities and is to be incorporated into Safe Operating Procedures (SOPs) when necessary. SOPs and/or equipment-specific shutdown procedures must individually address energized equipment and provide detailed procedures that must be followed for isolating energy sources.

B.   Performing a Lockout/Tagout

Employees are ultimately responsible for ensuring that equipment is disconnected from all hazardous energy sources and secured in a zero-energy state before attempting any work that could expose them to hazardous energy sources, including maintenance and servicing activities. Each exposed employee is responsible for applying the lockout device to secure the input energy source(s) prior to beginning any work activities. Whenever more than one employee is working on the same piece of equipment or project, a group lock box or multi-lock hasps (gang locks) shall be used and each employee shall apply his/her own lock. Another person's lockout shall never be relied on for protection. The following steps must be followed before work begins:

   i.   **Notify Equipment Users of Work**: Before beginning work on any equipment or system, always notify equipment users and employees in the area who could be affected by the shutdown. Whenever possible, work should be scheduled and coordinated with other staff members to minimize programmatic interruptions.

   ii.   **Reference Equipment-Specific Shutdown and LOCKOUT Procedures**: Facility-specific Safe Operating Procedures (SOPs) must identify and include lockout procedures for equipment used in the lab, to include, as a minimum, description and location of the energy-isolating device(s). Large and complex pieces of equipment are required to have written procedures which prescribe specific shutdown and lockout methods (see Appendix 11.2). These procedures must be referenced and followed prior to performing any work activities.

iii. **Identify all Internal and External Energy Sources**: Many pieces of equipment have more than one energy source that must be controlled. Written Lockout/Tagout procedures are required for all equipment/systems that have more than one energy source. All external energy sources such as the input electrical supply, compressed air lines, water supplies, etc. must be addressed. In addition, internal energy sources such as charged capacitors, batteries, wound springs, etc. must be identified. Schematic diagrams and operator manuals should be available and referenced for assistance in identifying input power requirements and internal energy sources. The SOP or specific Lockout/Tagout (LOTO) procedure shall be referenced.

iv. **Physically Disconnect and Isolate Energy Sources**: Once all energy sources have been identified, the next step is to physically disconnect and/or shut off the source(s) with appropriate energy-isolating devices and secure them in the off/disconnect position. It may be necessary to leave bleed valves open to prevent the reaccumulation of stored energy. The written LOTO procedure must specify how this is accomplished.

v. **Perform Initial Verification**: Before applying the lockout device, verifications must be made to ensure that the energy sources have indeed been disconnected. Never assume that the proper energy source has been disconnected because of your own or someone else's familiarity with the system. Depending on the system and the type of energy source, verifications can be made with voltmeters, pressure gauges, etc.

vi. **LOCKOUT the Energy-Isolating Device**: Each energy source must be locked out to prevent others from inadvertently reconnecting or re-energizing the equipment. Lockout devices must always be applied at the input power source and not at the control circuit. Remember that many pieces of equipment have more than one switch or switching method from which they can be turned on, thereby making control circuits poor places to apply lockout devices.

vii. **Exceptions for Cord and Plug Combinations**: Equipment that can be deenergized for service by unplugging it from its power source does not require LOTO, if the unplugged power cord is under the exclusive and immediate control of the person performing the work (i.e., the plug is within sight and within reach of the person performing the work and there is no potential for stored/residual energy). Lockout/Tagout is required for all other equipment.

viii. **TAGOUT the Energy-Isolating Device**: A "Danger Do Not Operate" tag shall be installed at the energy-isolating device(s). The tag shall be marked with the name and phone number of the person performing the Lockout/Tagout, the date, and other necessary and appropriate information.

ix. **Perform Final and Periodic Verifications**: Always perform a final verification before proceeding with the work that is to be performed. Verification shall include checking that electrical systems show no voltage present (and are grounded if applicable); steam, fluid, and pneumatic systems are depressurized and vented (or drained if applicable); and all isolation devices are properly positioned, inoperable, and appropriately tagged. If the work will be performed for an extended time period, periodic verifications must be done to ensure the integrity of the lockouts that have been applied.

C. Performing a Tagout Only
   i. Some devices cannot be locked out physically. For these situations, a tag is applied at the energy-isolating device and at the control panel. When working under a Tagout only, a watch shall be posted if the energy-isolating device is not in the direct line of sight of the individual performing the work.
   ii. Additional measures are necessary to provide the equivalent level of safety available from the use of a lockout device. This may include removal of an isolating circuit element, blocking of a controlling switch, opening of an extra disconnecting device, or the removal of a valve handle to reduce the likelihood of inadvertent energization.
   iii. A Safe Work Permit shall be obtained from the ES&H Office when Tagout is applied to equipment or machinery in place of Lockout. The Safe Work Permit, issued in accordance with Woodlake Policy 4-10, Occupational Safety and Health (Contact Woodlake ES&H) shall verify and delineate the equivalent level of safety.

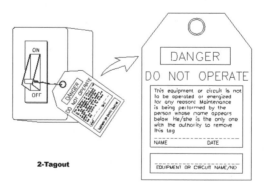

2-Tagout

180

iv.   If the energy-isolating device(s) on machinery or equipment is not capable of being locked out, that equipment shall be renovated or modified to accept a Lockout device.

D. Removal of a Lockout/Tagout Device

i.   **Routine Removal:** Lockout/Tagout devices shall only be removed by, or under the direction of, the individual who applied the device and whose name appears on the tag.

ii. **Non-Routine Removal**: If the person who applied the Lockout/Tagout device is unavailable, the team leader of the person who applied the Lockout/Tagout device has authority to remove the device, after receiving verbal approval from the individual who applied the lock, and ensuring that removal will not jeopardize the safety of other workers. As part of the removal and restoration process, the supervisor shall ensure that

- The equipment is operationally intact.

- Components within the Lockout/Tagout boundary that have been manipulated during maintenance, testing, or other work are positioned to permit operation, if required.

- Components that could cause automatic operation of a circuit breaker or a motor- or air-operated valve when control power or pressure is restored should be in a position such that automatic operation will NOT occur during removal of Lockout/Tagout.

- Upon completion of this verification process, the team leader shall cut the lock off. An entry shall be made in the Lockout/Tagout log. The individual who applied the lock shall sign in the log that he acknowledges that the lock was removed at the first opportunity when he returned to the job site.

E. Lockout/Tagout Hardware

Whenever possible, a padlock must be used to secure the disconnect mechanism or the energy-isolating device. The use of a padlock and appropriate key control measures is the most effective way to prevent other workers from inadvertently re-energizing a piece of equipment.

All new equipment installations and existing installations requiring periodic work involving hazardous energy sources shall be equipped with energy-isolating devices capable of being locked-out.

i.   **Padlocks:** All locks used for controlling hazardous energy sources shall be identifiably different from locks used for other applications, such as for securing gates or cabinets. Therefore, only padlocks that are red in color will be issued for the purpose of locking out hazardous energy sources. Likewise, red padlocks shall not be used for purposes other than controlling hazardous energy sources. Each padlock will be individually keyed and numbered. Locks are assigned on an individual basis. Sharing of locks or use of locks by anyone other than the individual to whom the lock was assigned is strictly prohibited. **Locking devices must be substantial so that they cannot be removed or bypassed while workers are depending on them for protection.**

Locks used for controlling hazardous energy sources must be accompanied by a "Danger Do Not Operate Tag".

ii. **Key Control:** Keys for locks, which are used to control hazardous energy sources, must be strictly controlled for the program to be effective. Each padlock will have only one key. The key will be assigned to the employee at the time the lock is checked out.

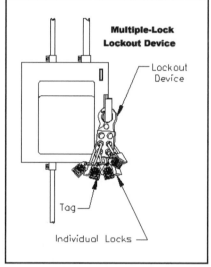

iii. **Multi-lock Hasps (Gang Locks):** Whenever more than one employee is working on a system or piece of equipment, multi-lock hasps or a group lock box must be used, and each worker must apply his/her lock to the hasp. Multi-lock hasps are available from commercial suppliers or the ES&H Lockout/Tagout Coordinator for short-term or emergency use.

iv. **Group Lock Box:** A lock box that permits each employee to place his/her personal lock on the box containing keys to locks on equipment being repaired or serviced. Individuals place one lock each on the lock box to effectively secure multiple sources of energy. As each person finishes his portion of the work, that person removes his lock from outside the lock box. When the last lock is removed, the keys to re-energize the equipment are accessible. Use of a Group Lock Box reduces the number of locks required to lockout a piece of equipment having multiple energy sources.

v. **Supplemental Locking Devices:** A number of energy-isolating mechanisms such as circuit breakers, valves, plugs, etc. may require the use of supplemental equipment before a padlock can be physically applied for the purpose of securing an energy source. Devices are now available which allow workers to lockout energy sources that previously could not be locked out. Whenever possible, supplemental devices must be obtained and utilized in lieu of simply applying a tagout. These devices are available from commercial suppliers, or from the ES&H Lockout/Tagout Coordinator for short-term emergency use.

vi. **"Danger Do Not Operate" Tags:** Customized Woodlake "Danger Do Not Operate" tags are available at the Hazard Communication stations located in Building 16, and from ES&H Office Staff. At a minimum, tags must include the following information: the name and telephone number of the individual who is applying the tag, and the date that the tag is applied. It is also helpful to include additional information on the tag as to its purpose.

"Danger Do Not Operate" tags must never be used on equipment that is energized or in service. Tags used for the purpose of controlling a hazardous energy source shall be durable enough to withstand exposure to the environment where they are installed, for as long as they are expected to be there.

F.   Working on Energized Equipment

Because of the high level of hazard, work on energized equipment shall be strenuously avoided. Equipment shall be designed and installed in a manner that eliminates the need for energized work, or the work shall be performed outside normal business hours when the equipment can be deenergized, if possible.

Situations may still occur that require work to be performed on equipment that is not in a Zero-Energy-State, such as measurements, adjustments, calibration, or trouble shooting. In these instances, necessary controls shall be identified and implemented to reduce the risk to an acceptable level. These may include, but are not limited to having a properly trained and equipped observer present during the work, using appropriate personal protective equipment, and limiting the volume or strength of the energy source.

i.   Work on energized equipment that must be done on a periodic or routine basis shall be addressed in an approved SOP, or other formal procedure.

ii. A Safe Work Permit shall be obtained from the ES&H Office for work on energized electrical equipment that falls outside the realm of an SOP.

iii. The definition and specific requirements for work on energized electrical systems are included in the Woodlake Electrical Safety Program. Within the requirements of that program, certain types of energized electrical work may be done without a SOP or Safe Work Permit.

G.   Training, Qualifications and Program Assistance

i.   All new employees will receive basic training on Woodlake's Lockout/Tagout Program, and how it may affect them, during New Employee ES&H Orientation.

    ii. All employees who perform work on power operated equipment that could be energized are required to complete Lockout/Tagout training. The Woodlake ES&H Office, as scheduled by Human Resources, will provide the training. The course will be conducted in accordance with the Course Outline on file with Human Resources to ensure all regulatory requirements are met. Only those employees who have completed Lockout/Tagout training will be allowed to checkout locks and supplemental equipment. Records maintained by Human Resources will be used for verification of training.

    iii. Never attempt to work on equipment if there is uncertainty about whether all energy sources have been identified and secured. For assistance in identifying energy sources and/or applying Lockout/Tagout devices, contact coworkers who are familiar with the equipment, the ES&H Office or Site Operations, and the written Lockout/Tagout procedure for the specific equipment/system.

H. Program Review

An annual review of the program shall be performed to ensure that program elements are in place, employees are properly trained, and program implementation is being documented. The ES&H LOTO Coordinator shall conduct the review. The review is structured to detect and correct all deviations or inadequacies observed. The Woodlake Lockout/Tagout Program Exception Log will be used to document this process.

APPENDIX 11.2

(Sample)
Equipment Specific Shutdown/Lockout Procedures
Digesters D-001 through D-010

This procedure shall be followed to deenergize LOTO Digesters D-001 through D-010. This procedure must be completed prior to performing any repair, service, or maintenance to the equipment.

1. Steam:
    - Turn off steam at valve SS-015; apply clamshell Lockout/Tagout (LOTO) device, a properly completed Woodlake LOTO tag and lock.
    - Open bleed valve BV-015 to dissipate accumulated steam.
    - Leave valve in "open" position and apply a clamshell LOTO device to valve. (This will prevent the reaccumulation of steam energy.)
    - Apply a properly completed Woodlake LOTO tag and lock.
    - Turn "off" the digester jacket steam supply at valve SJ-015.
    - Apply clamshell LOTO device and a properly completed Woodlake LOTO tag and lock.
    - "Open" bleed valve BV-016 to dissipate residual pressure. Leave LOTO BV-016 in the "open" position.
2. Electrical:
    - Deenergize the agitator motor by turning "off" disconnect switch MC-015.
    - Apply a properly completed Woodlake LOTO tag and lock.
    - Turn "off" the screw feed conveyor at disconnect switch FC-15.
    - Apply a properly completed Woodlake LOTO tag and lock.
    - Turn "off" power to the clamshell heaters at breaker PB-15, breakers 23, 24, and 25.
    - Apply breaker lockout devices and a properly completed Woodlake LOTO tag and lock.
    - Operate control switches or use a multimeter prior to beginning work to verify equipment is in zero-energy state.
3. Compressed Gas/High Pressure:
    - Turn "off" the flow of nitrogen at the nitrogen gas cylinder located adjacent to the digester bank.
    - Apply clamshell LOTO device and a properly completed Woodlake LOTO tag and lock.

- Open bleed valve BV-01 and release accumulated pressure.
- Leave valve open and apply a clamshell LOTO device and a properly completed Woodlake LOTO tag and lock.

4. Conduct Work
5. Return equipment to proper and safe operating condition
6. Replace guards, interlocks, and other equipment
7. Remove LOTO locks and devices
8. Close bleed valves
9. Return operating controls to "normal operating" position

Form 11.1
Hot Work/Open Flame Permit

_____ Line Break _____ Hot Work _____ General Work

Project Coordinator   Work Area

Work to be Performed

Effective Date   From AM/PM to AM/PM

AREA NOTIFICATION (REQUIRED – PRIOR TO CONTINUING)

Area Supervisor or Rep. Date _____ Time_____

GENERAL WORK _____   ### LINE BREAKING _____

1. Area Notification of Equipment Status____   1. Fire Extinguisher Present ____
2. Barricade Present   ____   2. Barricade Present   ____
3. Test Nearest Eye Wash & Shower   ____   3. Specify Required PPE   ____

HOT WORK/BURN PERMIT _____

1. Measure combustible gas content (Acceptable range 0%, Indicate % /   ___%
    Meter No.)   # ___
2. Manholes, doors, windows, and sewer connections covered or protected   ___
    (if possible)
3. ABC fire extinguisher present   ___
4. Sprinkler protection system is operable (verify with security if necessary)   ___
5. Smoke Detectors disabled by Fire Protection Engineering in immediate area   ___
6. Fire watch (required if combustible material is located with 100 foot radius) ___
7. All combustibles moved at least 50 feet from the work site   ___
8. Area flagged off and/or welder's curtain in place   ___

APPROVALS REQUIRED PRIOR TO BEGINNING WORK
"OWNER" EMPLOYEES – SIGNATURE REQUIRED

Contract Coordinator Date _____ Time _____

Safety Coordinator Date _____ Time _____

Fire Inspector Date _____ Time _____

I certify that all safety factors have been considered and authorize the work to begin.

PERSONNEL PERFORMING THE WORK – SIGNATURE REQUIRED

Contractor Employee Date _____ Time _____
Responsible Supervisor Date _____ Time _____
I certify that all safety factors have been considered and that I will follow the permit
requirements.

WORK COMPLETE — AREA CLEAN and RESTORED
TO SAFE CONDITION

Owner Employee _____ Person Performing Work _____

Job Completion: Return this permit upon job completion to the approving area
supervisor. The area supervisor will write "Complete", date and initial across the
face of the permit and forward to the Safety Department.

Chapter 12

Evaluation of Safety Program Effectiveness

I.   Introduction

Ensuring that safety is being integrated into contracted work practices is an important element of project management. It is in the job-site manager's interest[1] to ensure that this is being done to avoid the uninviting prospect of the occurrence of personal injuries that may lead to project delays and potential claims or lawsuits. In fact, each employer in the contracted work hierarchy has a vested interest in making the same assurance for the same reasons.

Just as the site host wishes to know that his CM is implementing an effective project safety program, the CM should verify that the general contractor is implementing an effective safety program, as specified in the contract. In turn, the general contractor should ensure that his sub-tier contractors are controlling workplace hazards and are integrating safety into their daily work practices.

---

[1]   OSHA 29 CFR 1926.20 General Safety and Health Provisions
ANSI A-10.33 (1992) Safety and Health Program Requirements for Multi-Employer Projects
ANSI A-10.38 (1991) Basic Elements of an Employer Program to Provide a Safe and Healthful Work Environment

Finally, the owners of sub-tier contractors should ensure that their on-site personnel are controlling the hazards identified during work planning and that their workers are working safely.

The first challenge faced by most project managers in regard to evaluating safety program effectiveness is determining what should be assessed and what measures of performance should be used. There are no hard and fast rules regarding how evaluations should be conducted nor is there a single most effective measurement system or perfect measure of safety performance. It is only through a systematic evaluation of a variety of elements that an accurate assessment of the degree to which safety is being integrated into job-site work practices can be determined.

II.   Evaluating Safety Program Effectiveness

Job-site managers who monitor safety performance and require that field inspections be performed regularly enjoy above average safety performance from their subcontractors. Roger Liska of the Construction Industry Institute[2] compared the differences between projects with good safety records and those projects with excellent safety records. He found that projects considered to be "excellent," as measured by low injury rates, included the requirement that safety evaluations be conducted regularly. Excellent projects capitalize on the concept stated by organization theorist, Mason Haire: "What gets measured . . . gets done."

> "The simple act of putting a measure on something is tantamount to getting it done. It focuses management attention to it."
>
> Thomas J. Peters and
> Robert W. Waterman, Jr.
> *In Search of Excellence*[3]

A. Accident Statistics

Accident statistics are the most commonly used measures of safety performance. Occupational injury rates that are reported in terms of frequency and severity over a specific period of man-hours worked, provide a relative measure of the effectiveness of the safety program in place. Although injury statistics only provide a retrospective analysis of injury experience, they are an indicator of how effectively a safety program was managed.

Statistics do not provide an indication of which program safety elements were implemented most effectively or where program weaknesses may have resulted in injuries, but they do provide a means of comparing the results of a safety program against previous loss experience or the industry in general. This

---

[2] Liska, Roger W., David Goodloe, and Rana Sen: "Zero Accidents," *Source Development 86,* Construction Industry Institute, Austin, Texas, January 1993.

information also enables organizations to compare their safety performance with that of other similar organizations.

On long projects, statistics can be used to compare relative performance between contractors or to demonstrate the effectiveness of long-term safety initiatives. Additionally, incident rates can be used to evaluate the effectiveness of supervisors in regard to their ability to manage work safely.

1. Limitations

A fact that must be kept in mind when comparing incident rates is that they are subject to many influencing factors. Differences between the interpretation of an organization's reporting of occurrences, the effectiveness with which an organization captures information regarding occurrences, and the differences in risks of the work being performed by each organization will have a bearing on the numbers. Small differences in incidence rates may mean nothing more than the fact that there are these differences in who is included in the man-hour counts, how injuries are classified, and if all injuries are reported.

Measures of the consequences of injuries and equipment damage are generally reported[3] as

- lost work day case (frequency) rates
- lost work days (severity) rates
- medical treatment case rates
- occupational illness case rates

Injury frequency rates are based on the number of injuries that have occurred over the period of one year, relative to a work crew of 100 men. In other words, a work crew of 50 people that experienced 2 lost time cases in a year would have a lost time rate of 4 since that would be the number of accidents that could be expected from a crew twice the size (100 men) during that same year. Another measure of safety performance is the severity of the injuries, which is measured in terms of lost work days.

Incident rates are subject to many influencing factors. A few considerations that must be kept in mind when comparing incident rates include

- the effect of differing interpretations between organizations' reporting criteria,
- the effectiveness of the capture of information regarding occurrences,
- the inherent risk in the work being performed by each organization,
- the measurement base for the calculation of incident rates.

---

[3]   29 CFR PART 1904 Recording and Reporting Occupational Injuries and Illness.

B. Workplace Inspections

Workplace inspections are a more effective means of monitoring the health of a safety program than relying only on statistics. Through inspections, the job-site manager can determine the degree of influence being exercised over site conditions, the control of hazards, the enforcement of safety standards, the use of required personal protective equipment, and the degree to which safe work practices are being applied.

The convention on construction projects is to dispatch safety inspectors to rove the job-site to identify instances of failure to comply with Occupational Safety and Health Act regulations. This approach has serious drawbacks. Most projects don't have the resources to provide sufficient inspectors with the proper training to maintain an effective level of vigilance. It also results in a random and often skewed perspective of job-site conditions. While this may be the only way OSHA inspectors can adequately sample the many construction sites they are expected to monitor, it is not an effective manner to collect job-site safety performance information and to promote a proactive safe work environment. In fact, designating inspectors to monitor workplace safety may only motivate workers to spend part of their effort being on the lookout for these safety vigilantes to ensure they are not caught working; for surely the inspector will find something wrong with that they are doing.

A more effective approach of evaluating job-site safety implementation is to enlist the support of project management and the workforce. This is a departure from conventional practice, but facility and contracted workers generally know where the unsafe work conditions are and can identify unsafe work practices. They will make constructive observations if encouraged to do so and they see that their observations are acted upon.

C. Joint Inspections

To further expand on the concept of having inspections conducted by individuals other than designated inspectors is the concept of Joint Job-Site Inspections. Joint inspections enlist the involvement of client and construction manager representatives, the construction manager and general contractor representatives, and so on down through the organizational structure, to team together to conduct job-site inspections on a regular basis. This collaborative approach provides a forum for managers to actively promote safety as well as supplying them with first-hand knowledge of site conditions and the degree to which safety is being practiced.

Where the joint inspection team approach is adopted, each day a team of two individuals is scheduled to conduct an inspection, preferably at their convenience. They should be provided with a specific objective. This might be a geographic area or a subject matter. Every inspection does not need to cover the entire site, or the entire spectrum of safety considerations. One inspection could focus on compliance with electrical safety issues such as ground fault circuit interrupters.

They might be asked to look for material blocking access to breakers, the use of extension cords and housekeeping, etc. This will provide a focus for the team and help them maintain interest in the program. Technical specialists can be periodically invited to accompany the joint inspection teams. The safety professional and occupational health specialists can provide the team members with a technical perspective on the issues being evaluated and educate them with regard to specific safety issues with which they might not be familiar.

Safety checklists can also be provided to the joint inspection teams to initially guide them. Safety inspection checklists, when used, should be developed to reflect the specific issues of the job-site. Checklists are useful as an aide-memoir, but it must be clear that the list may not be all encompassing. A **Safety Inspection Checklist** (Form 12.1) is an example of the sort of information that can be provided to the inspection groups as they develop confidence in their ability to identify substandard safety conditions. Individuals con-ducting the inspections should be reminded not to feel constrained by the checklist. Other safety issues they may identify should be noted as well.

This condition will creep up on you if you are not vigilant.

Each inspection team should be requested to submit a report of their observations and what they did to correct identified substandard conditions. They should also identify those items they were not able to correct and that require follow-up action. Observations made during inspections should be recorded in a systematic format such as the **Safety/Health Inspection Action Form** (Form 12.2). The inspectors should be specific regarding what was determined to be deficient, where the observation was made, and who should be responsible for the follow-up corrective action. Where imminently dangerous situations are identified, activities should be stopped until corrected.

Inspections are an effective means of demonstrating support of the safety program and provide a good opportunity to reinforce examples of good work practices and performances identified during the inspection process. In addition to demonstrating partnership in the safety process to all project personnel, this approach provides the opportunity for individuals throughout the management and supervisory hierarchy to better understand the challenges and expectations faced by each other in regard to integrating safety into daily work practices.

Form 12.1
Safety Inspection Checklist

Area Inspected: _____ Date: _____

Persons Conducting Inspection: _____

Note:
1. Place a "✓" mark in appropriate column
2. All "No" answers should be explained or result in suggestions
3. Line through "OK" and "NO" boxes when not applicable.

OK	NO	#	Ref
			Housekeeping
☐	☐	___	Walkways clear of litter
☐	☐	___	Lumber stripped of nails
☐	☐	___	Proper lighting
☐	☐	___	Control of material & tools
☐	☐	___	Roadways & walkways clear
			Fire Extinguishers
☐	☐	___	Conspicuously located
☐	☐	___	Suitable distribution & size
☐	☐	___	Yearly recharge & tags
☐	☐	___	Clean & properly mounted
			Means of Access and Egress
☐	☐	___	Aisles free of obstructions
☐	☐	___	Exit signs visible
☐	☐	___	Proper lighting
☐	☐	___	Ladders in use are secure
☐	☐	___	Stairs free of obstructions
			Job-Site Premises
☐	☐	___	Open edges guarded
☐	☐	___	Floor hole openings covered
☐	☐	___	Overhead hazards barricaded

OK	NO	#	Ref
☐	☐	___	Floor hole covers secure
☐	☐	___	Rebar covered to prevent impalement
☐	☐	___	Hazards protected from children
☐	☐	___	Exit routes clearly posted
			Crane Operations
☐	☐	___	Safety latch on chain hook
☐	☐	___	Anti-two blocking device in place
			Mobile Equipment
☐	☐	___	Proper loading of equipment
☐	☐	___	Loads properly handled
☐	☐	___	Overhead obstructions flagged
☐	☐	___	Equipment properly parked
☐	☐	___	Roll-over protection as needed
☐	☐	___	Reverse alarm operating
☐	☐	___	Seat belts in use
			Air Compressors/Hydraulic
☐	☐	___	Equipped with pressure gauges
☐	☐	___	Hoses & regulators in good condition
☐	☐	___	Safety valve provided
☐	☐	___	Hoses secured to fittings

Form 12.2

Safety/Health Inspection Action Form

Location Inspected: Bagging Plant

Date & Time of Inspection: 11/12/99 11:00 a.m.

Inspector and Title: Richard Hislop, Safety Manager

Date of Report: 11/13/99

Pending Items (Marked with *)

Ref No.	Observation/ Hazard Description (Define Specific Location)	Corrective Action Recommended	Person Responsible	Hazard Classification Date Completed	Corrective Action Taken
1	Handrail missing on gallery second floor by column 32.	Contractor XYZ directed to correct condition before further work proceeds.	Bob Jones XYZ Foreman	**Classification: A** 11/12	XYZ replaced the handrail by the end of the same work day.
2	Painter was painting from a folded A–frame ladder leaning against a wall.	The inappropriate use of the ladder was brought to the attention of the painter. He agreed not to do it again.	Glen Kailus PFS Supervisor	**Classification: B** 11/13	Supervisor reported that the safe use of ladders was addressed at the next pre-work briefing.
*3	Construction debris is around disposal bins.	Contractors to be advised of need to be more careful in getting debris into containers at next coordination meeting.	Bill Wiggins GC Super.	**Classification: C** Correct within one week.	
	Comments: (Additional comments on back) Signature:				

Hazard must be corrected by:

Class A – Corrected before work continues

Class B – Corrected within 1-2 shifts

Class C – Corrected before defined date

195

D. Work Practice Evaluations

Growing in acceptance is the focus on evaluating craftsmen's work practices. It is generally accepted among safety professionals that the majority of workplace accidents are the result of unsafe behavior and work practices. Various percentages bandied about suggest that between 90 and 96% of personal injuries are the result of unsafe work behaviors. Therefore, it would make sense that endeavoring to identify unsafe behaviors and encouraging workers to follow safe work practices should result in fewer job-site accidents.

The work practice evaluation approach focuses on the manner in which workers perform their work to determine if it is being done safely. The greatest challenge in conducting these evaluations is to overcome the natural inclination of the observers to limit their focus to workplace <u>conditions</u> and compliance with regulatory standards rather than how work is being performed. Observers should focus on work in progress rather than job-site conditions. When work practices do not appear to be "right," observers must determine why work is being done that way. This approach also provides the observers the opportunity to commend individuals who are following good work practices. This is quite a bit different than the "Sea Gull" approach to safety that is often used in construction, where an inspector swoops in with a great flap. The inspectors identify something wrong, assume that the craftsman is working unsafely on purpose, assigns blame, issues an authoritarian directive to stop that practice and then flaps off to harass another worker.

The process of approaching workers during an inspection is a technique that must be learned by most individuals. In the work practice observation process an important fact must be recognized, individuals work the way they do because they either learned to do it that way, or they feel it is expected of them. Few individuals work with the express objective of injuring themselves.

In order to make a lasting correction, the observer must first determine the source of the problem, and that requires approaching the individual. Generally, the reason for inappropriate/unsafe work practices is the result of some pre-existing condition or constraints that give the worker little choice. These constraints can generally be sorted into either personal factors or job-related factors.

1. Personal factors

Personal factors consist of lack of craftsman skill/knowledge or lack of motivation to perform the work properly. If the worker does not know how to perform the work properly/safely, that individual's employer should remedy the situation by educating him or replacing him. If the cause of the unacceptable behavior is the lack of motivation to work safely, the source of this problem may be that the worker perceives that doing the job exactly according to procedures is not important. The individual may be convinced that proper work practices take more time and require more effort than the way he has been doing it. Finally, it may be that short-cutting standard procedures is rewarded or is perceived as appreciated based on the positive feedback the worker gets from his supervisor for getting the job done quickly.

2. Job factors

Job factors might possibly play a role in the reason a worker is not following safe work practices. It could be the absence of operational procedures, inadequate operational procedures, or inadequate work standards that provide the worker with guidance how to do the work properly. It might be inadequate communication of expectations regarding procedures or work standards, or it might be that inadequate tools or equipment compel the employee to use inappropriate tools.

E. Evaluating Employee Behavior

When conducting employee behavior observations, the process begins as soon as the observer enters the work area. Start with an evaluation of the immediate reactions of craftsmen as soon as the observer enters the work area. Focus on the actions of employees when they perceive that they are being observed. As soon as individuals think that someone might be evaluating their safety performance, they assess what they are doing and attempt to correct what they recognize to be unsafe. They will adjust their personal protective equipment, safety glasses, machine guards, etc. Another immediate response is to discontinue working or turn to another task. The craftsmen may rearrange the work they are doing to eliminate the unsafe practices they recognize or change the position they are working in if it is improper. Obviously these adjustments soon stop. If these activities are not noted as they occur they will be lost to the observer, as they will not be repeated.

Personal protective equipment (PPE) requirements are generally self-evident. Most construction environments require craftsmen to wear hardhats, safety glasses, and sturdy work boots. However, the need for other protection may not be standardized for all employees. Where an employee is observed not wearing required PPE or there is some indication that PPE might be required but is not being used, there is an opportunity to approach the individual to determine the reason. For example, a welder is working chest high or over his head in a short sleeve shirt. There is the real potential for burns. If the observer is uncomfortable approaching the worker, the individual's supervisor should be directed to approach the worker.

Positions of individuals as they perform work are key factors in the occurrence of personal injuries. Thus, the potential for falling, being pinned between two objects or being struck by moving objects should be evaluated.

Equipment and tools that are not proper for the job, or have not been maintained with care can result in personal injuries. The failure to use proper tools may be the result of the employee trying to get the job done quickly and not being willing to return to the tool crib out of a misguided sense of work priorities.

Procedures and Housekeeping are also significant elements of a safe job site. There are those who will argue that housekeeping takes time away from production. However, in my experience there is a direct correlation between good housekeeping and low injury rates. Supervisors will argue that few if any injuries are the direct result of poor housekeeping. However, the fact is that where there is good housekeeping there are few injuries from any source.

Those performing workplace observations should use a "Sampling Worksheet". This information should be compiled in a "Safety Sampling Report" and distributed at least monthly. With time, the elements most in compliance should be replaced with other issues that require more attention.

F. Evaluation of Documented Safety Programs

The evaluation approach is one that would be used by a very involved owner to judge the effectiveness with which a construction manager or general contractor is implementing his documented safety program.

Like financial audits, safety program audits typically address clearly defined criteria. In the construction industry packaged audits have become common place. Entities peddling packaged audits advocate that specific program elements must be included in the evaluation and high ratings can only be achieved if those elements are present. Some advocate five critical elements, others nineteen, and some thirty. The basis for their respective claims of success is that their clients have reduced their accident rates through the implementation of their particular program elements. The fact is that safety performance improves following regular audits because greater attention is focused on safe work practices. The improvement in performance might be nothing more than an improvement as a result of the "Hawthorn Effect".[4] Their program elements are not necessarily the reasons for the reduction in losses. Their previous clients might have had a rigorous safety program in place. This is communicating a message to the employees that management is focusing on some specific facet of their work.

Comprehensive assessments should be conducted using a standard audit protocol to ensure consistency with successive audits. Written guidelines for interpretation and evaluation of the protocol questions should be used to guide the auditor(s) in making consistent applications. Those persons conducting the audits should be trained in safety and health program auditing techniques. While the job-site manager can certainly conduct the assessments using his or her own safety and health staff, an independent safety professional, approved by both parties, might be viewed as being unbiased. The result of an audit should be a report highlighting the strengths and positive aspects of the contractors' programs as well as identifying areas where improvements are needed and making specific recommendations for such improvements. When an audit is conducted, its objective should be clearly

---

[4] Hawthorn Effect. This is the affect-attributed change in behavior as the result of surveys or studies done on individual's behavior. The Hawthorn Effect drew its name from a study conducted at the Western Electric Hawthorn plant in Cicero, Illinois, where the effect was first noticed in the 1930s. Researchers tried to measure how changes in environment (levels of lighting) affected worker productivity. They found that productivity improved in all test groups, regardless of whether the light level was raised or lowered. The conclusion was that just being part of a study or receiving individual attention motivated workers to perform better.

defined. The auditors should identify the specific questions to be addressed. This will assist in defining the method by which the audit is to be conducted, the documentation to be reviewed, and individuals to be contacted. The contractor to be audited should be advised of the audit, its purpose, the information required for the review and the name or positions of individuals to be interviewed. The contractor should also be given a general indication of the length of time and any specific requirements of the audit. Finally, the contractor should be advised when they will be debriefed and provided with an audit report.

Job-site safety evaluations should be structured to determine if the following elements are included in the safety program structure:

1. Line management is responsible and held accountable for safety performance.
    - Accountability
2. Clear roles and responsibilities for safety have been defined.
    - Responsibility matrices
    - Technical and safety requirements of supervisory staff
3. Both crafts people and management have the requisite knowledge, commensurate with their responsibilities, to recognize and respond to safety-related issues.
    - Training content relative to job-site hazards
    - Documentation of training verification (i.e., Exams)
    - Number of orientations versus number of employees
4. Priorities between safety, quality and schedule are balanced.
    - Facility work planning
    - Contractor work planning
5. Relevant safety standards and requirements have been identified and communicated as appropriate to the workforce and line management.
    - Directives
    - Oversight of compliance with standards
    - Type of equipment inspected
    - How equipment to be inspected is identified
    - Elements included in inspection checklists
    - Quality of observations
    - Measures taken where defects are identified
6. Hazard controls are tailored to work being performed.
    - Procedures developed for high-risk tasks
    - Review and approval procedures
    - Personal protective equipment, availability and use where required
    - Stop work authority

7. The work process is structured such that there is
    - Clear definition of task/work scope.
    - Work task planning includes analysis of hazards.
    - Controls of identified hazards are developed and resources are available to control the hazards.
    - Work direction and oversight ensure that work is performed safely.
        ◊ Job-site surveillance and inspections
            - Frequency
            - Documentation
            - Quality of observations
            - Process by which observations/hazards are reported
            - Closure of corrected items
    - Means are in place to provide feedback to management and the workforce to continuously improve working practices.
        ◊ Supervisor performance evaluations
        ◊ Tracking, trending and lessons learned
        ◊ Accident/incident investigation process
            - Quality of investigations and documentation
            - Number of investigations versus number of incidents

Following an audit the contractor should be required to develop an action plan for bringing deficiencies into compliance with contract requirements. The contractor should adhere to this plan until all requirements are met. The assessment results, the report, and the remedial action records should be maintained as a part of the contractor's file to be considered in future bids/evaluations.

III.   Frequency of Evaluations and Audits

The frequency of inspections and audits is contingent on where in the organizational hierarchy the evaluation is initiated.

Job-site inspections should be conducted regularly. The size of the project, the inherent hazards and the number of individuals working on the project will have a bearing on the frequency. A joint owner/job-site manager inspection might be conducted every other week, while the job-site manager and general contractors walk the site weekly. Superintendents should inspect work daily and foremen should include safety evaluations in their routine interfaces with their employees several times a day.

Work practice observations should be conducted by supervisory personnel and peers on a regular basis, depending on their position within the organizational hierarchy. A manager might conduct one observation each week, while a foreman might be expected to conduct an observation several times a week.

A formal audit of a contractor's safety program will generally only take place on projects with a duration of a year or greater. The evaluation of contractors on site for only a few weeks or months will be limited to their job-specific procedures and JSAs.

IV.   What Should Be Done With Evaluation Results?

The manager of each organizational unit assessed should be apprised of evaluation findings and recommendations. The responsible manager should be expected to develop corrective actions and establish dates by which corrective actions will be completed.

Recommendations made by evaluators should be guided by the objective of eliminating workplace hazards. The individual responsible for taking the corrective action should also be identified in the report. This avoids any confusion that might arise if someone is not specifically allocated the responsibility for taking action. Do not accept a narrative report that only describes the program and insist that the difficult issues be addressed. A good evaluation should lead to program improvements.

V.   Summary

During the selection process each contractor should be made aware of the job-site manager's intention to perform periodic safety assessments. The progressive contractor will recognize this as an opportunity to develop his safety program while utilizing the resources of the proprietor to conduct the assessments. Participants in a well-organized evaluation process should come to recognize that the procedure will not only identify safety program strengths and weaknesses, but also provides a forum to improve management contractor relationships.

Chapter 13

Safety Statistics
and Reports

Everyone wants to know the score. "How well are we doing?" "Are we doing okay?" This is also the case regarding safety in contracted work. However, the challenge faced by most individuals asked to answer this question is determining what should be measured and how to best report that information.

I.   Introduction

Philosophies and opinions regarding what safety information should be measured and what should be reported are as abundant as there are individuals. Some individuals feel that accident statistics are ineffective measures of safety performance. They argue that statistics provide only a limited view of safety performance and, as they offer no useful information regarding the causes of losses, therefore they are not an effective management tool.

Both baseball and football seem to effectively use game scores and statistics as measures of outcome and performance. However, both baseball and football rely on a variety of other performance measures for those interested in the technical details of the game, such as individual batting averages, home runs, or quarterback sacks and yards gained by specific individuals. With safety, there are similar quantitative and qualitative measures of performance.

Quantitative safety information can include the number of activities that have taken place, e.g., 16 safety inspections conducted this month, 5 safety violations reported, no lost time cases, etc. Qualitative safety information reflecting relative performance can also be developed, e.g., superintendent Jack Jones' workers experienced 15% fewer injuries last month, which was a greater improvement than the 6% decrease in project injuries over the same period.

II. What Should Be Reported?

If you are tasked to produce a safety report, your first order of business should be to determine to what specific reporting requirements you are bound by contract and by law. OSHA requires that each employer maintain an OSHA 200 Log. The contract may specify that monthly man-hour reports and recordable cases be reported to the host employer. In fact, the contract may define how this information is to be presented.

Owners wish to be assured that an effective safety program is in place and they want safety information to convince their Board of Directors that they are being good corporate citizens. As the job-site manager or field engineer you will want measures that will enable you to judge the effectiveness with which the general and prime contractors are implementing their respective safety responsibilities. The general and prime contractors would probably like to know how effective their supervisory personnel are with regard to managing the safety of their respective work crews.

The information to be included in your safety report may include the following:

Owner requested information
    Recordable rate
    Severity rate

OSHA reporting requirement
    OSHA 200 Log (lost time, severity and recordable cases)

Construction manager measures of performance
    Numbers of inspections conducted
    Number and type of unsafe work conditions observed

Contractor performance measurements (by supervisor)
    Injury statistics (lost time and recordable rates)
    Dollar value of personnel injury losses by supervisor
    Dollar value of equipment damage by supervisor
    Percentage of safe behaviors

The challenge is how to obtain this information and to present it in a manner that will effectively convey your intended message.

What Gets Measured Gets Done!

III.   Collecting Safety Information

Now that you have defined your report requirements you must establish a mechanism to collect this information. Fortunately, you specified in the contract that each contractor, along with his requests for payments, submit a Man-Hour and Incident Report (Form 13-1).

Additionally, every personal injury and near-miss case will be reported using the Immediate Occurrence/Near Miss Report (Form 13-2) which will give you a cross check on the Man-hour and Incident Report information.

The joint inspections and worker observation reports provide another source of information for your report.

IV.   Safety Report

The objective of the safety report is to raise the awareness of the project members, from the host employer to the craftsmen, of the status of safety integration into work practices. We have seen that each hierarchical level within the project needs information presented in a slightly different context.

A. Owners

The owners funding the contracted work effort generally want to be assured that work on their site is being performed safely and in compliance with regulatory requirements. To demonstrate this, the report should reflect the contractor's safety performance against recognized benchmarks.

Example: Lost Time Incidence Rates

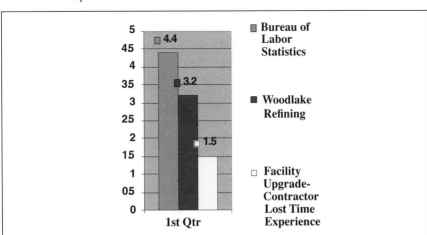

Figure 13.1 Loss rates as compared to some recognized basis

A graph that reflects relative performance between general industry, the corporate facility where the work is being done and the contractor incidence rates is an effective means of communicating this information. Well, it is good as long as the comparison is favorable. It is an effective and generally well-received report element.

Although statistics are not difficult to understand if you have learned how to interpret them, not everyone is comfortable with them. It is generally a good idea to include the same information in a non-statistical format, such as the example below:

Contractor	Lost Time Injuries This Month	Lost Time Injuries to Date	Manhours This Month	Manhours to Date
Electrical Svc.	0	0	1,600	7,500
Inter-Piping	1	1	3,400	9,200
Concrete & Forms	0	1	500	5,500

1.   Incident Rate Determination

Accident or incident rates are the most frequently used measures of safety performance. Incident rates provide a relative measure of occupational injuries expressed in terms of frequency and severity as experienced over a specific period of man-hours worked. They are calculated based on the generally accepted OSHA convention that reflects the number of injuries that have been sustained by an average work crew of 100 individuals over the period of a year. The number 200,000 represents the hours worked by 100 employees, working 40 hours a week for 50 weeks a year.

$$\text{Lost Time Incidence Rate} = \frac{\text{Number of Lost Time Incidents} \times 200,000}{\text{Number of Manhours Worked}}$$

Example:

If an organization experienced 3 injuries resulting in time lost over a period when 400,000 hours were worked, the incidence rate would be computed as follows:

$$\text{Lost Time Incidence Rate} = \frac{3 \text{ lost time injuries} \times 200,000}{400,000 \text{ employee hours worked}}$$

Lost Time Incidence Rate = 1.2

This formula is used to calculate the following rates:

- Lost Work Days Case Rate (Incident Rate)
- Lost Work Days Rate (Frequency Rate)
- Medical Treatment Case Rate
- Occupational Illness Case Rate

2. Relative Performance

Information regarding Lost Time Rates in general industry can be obtained from the National Safety Council (http://www.nsc.org/) or from the Bureau of Labor Statistics (http://www.bls.gov/oshhome.htm).

Their addresses are

National Safety Council

1121 Spring Lake Drive,

Itasca, IL 60143-3201

Tel: (630) 285-1121

Fax: (630) 285-1315

U.S. Department of Labor

2 Massachusetts Ave., NE

Washington, D.C. 20212

Tel: (202) 606-6170

Fax: (202) 606-6196

3. What Are the Problem Areas?

Attempt to define in the report what the sources of safety exposures might be. Where it makes sense, identify your assessment of the potential exposure each issue presents to the operation.

> *Example:*
> *"During the past month workplace observations identified the following distribution of sub-standard work practices:*
> - *poor housekeeping*   *33%*
> - *failure to follow JSAs*   *15%*
> - *inadequate marking of work areas*   *15%*
> - *failure to use PPE*   *10%*
> - *etc.*
>
> *Contractors are failing to dispose of waste at the end of each work day since the dumpsters are not being emptied with sufficient regularity. This issue was corrected during the last week of the month and housekeeping should no longer be an issue.*
>
> *Several instances of flammable liquids being stored in plastic containers not designed for the function and which present fire hazards. A safety notice was sent to all contractors to discontinue this practice."*

Owners want to know that a process is in place to identify program weaknesses and what is being done to correct them. Are the issues identified the result of difficulties in the implementation of the safety program? What support is needed, if any?

B. Job-Site Manager

The job-site manager, who has a legal duty to provide instruction and orientation to contractors regarding hazards familiar to the host employer and ensure that the contractors are exercising their responsibility as employers by implementing their own safety programs, needs information to judge the effectiveness with which both of these issues are being executed.

1. *Safety Program Implementation*
    When the safety program was defined, orientations were to be conducted and inspections were to be performed. A table that quantifies expected performance against that actually accomplished provides some indication of the efficiency of the execution of the activities.

    Example:

Safety Program Element	Planned	Executed	Compliance
Orientations	78	78	100%
Supervisor Inspections	17	17	100%
Joint Inspections	40	40	100%

> "***Note:*** *The joint safety inspections have been focusing on last month's top three program deficiencies: PPE, housekeeping and use of GFCIs. No instances of failure to use GFCIs were identified this month.*"

C. General and Prime Contractors

These organizations should be evaluating their own safety performance and generating their own internal reports. It is the exception where this is the case. Since the job-site manager has a vested interest in promoting safety awareness among all contractors on site, information relevant to the contractors' performance should also be captured.

Reporting relative performance among contractors is a powerful motivational tool if, by chance, the contractors get a copy of the project safety report being forwarded to the client. The following is an extract from a project safety report:

> "*Good News: On April 16 the project completed 1,500,000 man-hours without a Lost Time case. The project has earned its fourth President's Award for Safety Excellence.*
> *MKI Superintendent with the best safety record for the month of April is*

> **Arlie Behrens** – *and his crew*
> *Ricardo Medrano   Miguel Castro   Luis Epiayu   Inqui Epiayu*
> *Vicente Epiayu   Jose Gil   Teodulo Ceron   Francisco Perez*

> *Subcontractor with the best safety record for the month of April is*

Schrader Camargo and the crew from Components Rebuilding
Jose Casas, P.E.   *Manager*
Luis Sanches   *Oswalo Acuna*   *Javier Torregrosa*   *Jose Turriago*
Diaro Rangel   *Jose Perez*   *Rafael Ricardo"*

In this section a description of injuries and near-miss occurrences can be addressed along with lessons learned.

V.   Conclusion

There are no hard and fast rules about what should be in a safety report. The contents depend to a significant extent on the requirements of the host employer, the job-site manager, and what message project management wishes to convey to the project team members. Keep the message clearly in mind and structure the report to communicate it.

Form 13.1 Manhour and Incident Report

Project No.:

Contractor:

Report Period From: _____ To: _____

	Period	Cumulative	Comments
Manhours Worked			
Recordable Cases			
Lost Work Day Cases			
Lost Work Days			
Other Safety Incidents			

Remarks & Notes (Safety incidents, etc.)

Submitted by: Date: _____

Distribution by Contractor:

Procurement Manager

Project Manager

Safety Manager

Form 13.2
Immediate Occurrence/Near Miss Report

This report is based on information collected at the scene of events that have a potentially negative effect on the efficient operation of the facility. The information is to be distributed as soon as reasonably possible following the event. (Fax within no more than 3 hours of occurrence.)

Location of occurrence: _____   Date/Time: _____

Brief description of occurrence: _____

_____

_____

_____

Immediate action taken: _____

_____

_____

Estimated impact (i.e., lost time, $ damage): _____

_____

Medical Department diagnosis: _____

Transported to hospital No ☐ Yes ☐   Hospital name: _____

Injured employee: _____   Direct Hire Employee Y Contractor Y

Organization:_____   (Contractor's Employer: _____)

Supervisor: _____

Individuals who responded to the scene: Photographs taken at scene

Yes ☐ No ☐

Safety Coordinator _____   (Sketch here if needed)

FD Incident Commander_____

Others _____

Distribution: Fax #

____ Chief Operations Officer   ###-####
____ Safety Director   ###-####
____ Construction Manager   ###-####
____ Safety Coordinator   _____
____ Other   _____

Prepared by: _____

Chapter 14

Emergency and Crisis Management

"My name is Bill Adams. I am the superintendent of Able Construction Company. As you are aware we have just experienced an accident that has apparently resulted in power interruption to the neighborhoods in this area and we have had two workers injured. Due to the rush associated with this emergency I do not have any detailed information at this time. Please give me 45 minutes to gather some facts for you. In the meantime, please stay in this safety area until I return."

Is this how your superintendent would respond when faced with the Media at the front gate of your project site? If it isn't, you may have a crisis developing at your facility. Every organization will some day, in some manner, experience a crisis. How well we do in a crisis situation will be determined by how well we have prepared to deal with it.[1]

---

[1] Janine Reid, *The Crisis Management Game, A Crisis Management Simulation,* Chicago, IL. February 1998.

I.   Introduction

A contractor severs a power line running through a project site interrupting service to a nearby subdivision and television station. A gas line is punctured while the foundations for a new maintenance facility are being excavated. A roofing contractor dumps solvent down the vent shaft of an occupied building. A tower crane collapses into a busy downtown street.

Every year emergencies such as these take their toll on construction projects and businesses . . . in lives and dollars. Although there is little we can do to avert natural disasters such as tornadoes and floods, by careful planning we can minimize the damage either man-made or natural disasters might do to our projects. Having an effective emergency and crisis action plan already in place when an incident occurs places the organization in a better position to control the situation. It will reduce the severity of the outcome, normal operations will resume more quickly, and there will be potentially less negative impact on your organization's reputation and finances.

Not only does it make good business sense, but it is also a regulatory requirement that employers with more than 10 employees have an Emergency Action Plan.[2] In some situations the plan may be as simple as calling the fire department or evacuating a building. At the other extreme, a complex plan may be required to restore utilities after they have been damaged.

A confidential survey of the nation's chief executive officers of the Fortune 500 revealed that a staggering 89% of those who responded agreed that a "crisis in business today is as inevitable as death and taxes." However, 50% of the respondents admitted that they had not prepared a crisis management plan. Of those companies that had experienced a crisis, 42% still did not have a crisis management plan in the event another crisis occurred.[3]

The Objectives of Emergency Action Planning are
- Employee protection
- Protection of the facility and equipment
- Communication
- Maintenance of vital functions
- Accelerate recovery

Numerous events in contracted work environments can be considered to be "emergencies." Events, which are themselves a crisis or which could degenerate into a crisis situation, include

1.   On-the-job fatalities
2.   On-the-job accidents requiring hospitalization
3.   Damage to utility lines

---

[2]   29 CFR 1910.38 & 1910.120
[3]   *Crisis Management*, Steven Fink, pg. 67, American Management Association, 1986.

4.   Highway accidents
5.   Equipment failures
6.   Theft/embezzlement
7.   Noise/dust pollution
8.   Sexual harassment
9.   Contractual disputes resulting in litigation
10.   Labor strike/work stoppages

To begin to prepare for these scenarios, anticipate what could go wrong. Identify potential problems that would constitute an emergency where you work, through a "what-if scenario."

As the number of contractors brought onto the job-site increases, the probability of there being an occurrence rises accordingly. The incident which may precipitate an emergency may be the lack of new employee familiarity with hazards in a new work environment or the failure of work crews to communicate differences in their work practices with each other. In certain environments, such as in a refinery or chemical plant, individuals without appropriate process knowledge could potentially precipitate a major crisis.

How about the contractor working on the roof of a hospital who attempted to dispose of his roof compound solvent by dumping it down a vent pipe? What was the chemical? Was it carcinogenic? How long will it linger in the building? You say nobody would do anything that stupid! You have a crisis on your hands.

To complicate this situation, increased media coverage over the past few years has resulted in extremely aggressive reporters, who in their pursuit of news may exacerbate what might otherwise be only a minor incident into a crisis situation. If reporters believe there is a story, they will make sure to get one. The number of factors which differentiate an incident that receives significant media attention from one that is never heard of by anyone other than those immediately involved may be no more than a reporter's erroneous preconception of a situation, reinforced by the lack of appropriate feedback from the organization experiencing the incident.

To successfully manage an emergency situation and to prevent it from degenerating into a crisis, one must be in control. Control is achieved through preparation. Companies in crisis situations from which they could have reasonably recovered if left to their own means have been driven out of business as the result of poorly managed public relations. What could prove fatal to your organization is your inability to communicate your side of the story succinctly and accurately. If all you had to worry about was dealing with the incident, that would be one thing. But, the arrival of the media, with its microphones, cameras, and reporters wanting details complicates the situation.

Many organizations hesitate to address the subject of emergency management planning or discuss how they would handle a crisis situation, feeling this is equivalent to admitting that they may be poorly organized or have a less-than-adequate safety program.

On the other hand, organizations with emergency and crisis management plans in place have recognized that even if they have an effective safety program there remains the possibility of undesirable occurrences degenerating into a crisis situation. The process of preparing to deal with these eventualities highlights preventive measures and information that should be included in safety training. Additionally, planning and preparation activities generate general awareness of issues and concerns and this awareness in itself may prevent some occurrences.

II.   What is Emergency Management?

Emergency management is the dynamic process of preparing for, responding to, and recovering from a critical situation . . . an emergency. An emergency is an unplanned event that results in death or significant injuries to employees, customers, or the public. It can also be an occurrence that disrupts or shuts down business operations, causes physical or environmental damage, or threatens the facility's financial standing or public image.

What you may not realize is that you probably avert situations every day of your business life that could possibly result in a crisis. You may not be aware of it because you effectively manage the situations as they arise as part of your regular routine. Have you ever noticed how often workplace problems rear their heads more often when you are not on site? You are able to resolve most of them quickly, once they have been brought to your attention. What if you were not around? Is there a possibility that they could degenerate into a major problem? What would happen if a newsworthy occurrence did receive the attention of the media? Would it be controlled in time to avert a crisis?

Clearly not all incidents degenerate into frontpage news. But, the degree to which one is prepared and the manner in which a situation is handled will determine the impact an event has on the organization. One can certainly hope for the best, but being prepared is better. Every business, irrespective of size, should be prepared to deal with incidents that may happen to them. The only difference between companies will be the scope of what is included in the planning considerations.

From a practical point of view a crisis situation may have several outcomes:

1. Escalation in intensity
2. Falling under close media or government scrutiny
3. Interfering with the normal operations
4. Jeopardizing the positive public image enjoyed by a company and its officers
5. Damage to a company's bottom line

If any of these conditions develop, the situation will likely take a turn for the worse. However, if a situation which runs the risk of escalating is caught and dealt with in time, it may dissipate and be resolved. Instead of becoming frontpage news, the media may never link your name to any adverse condition or crisis.

III. Developing an Emergency Management Plan

Most households are only minimally prepared for emergencies. Those of us who live in areas such as the east coast which experiences frequent and violent storms, the midwest which is subject to flooding and tornadoes, or the west coast where earthquakes occur, recognize that we may lose our electric power occasionally. It is not a question of if it will happen, but when it will happen next. Some of us have flashlights and spare batteries on hand in anticipation of the next occurrence. Preparedness depends to a great extent on our understanding of the hazards to which we are exposed and the potential losses associated with the hazards. Many of us who have experienced household emergencies can look back and identify a number of things we could have done to have minimized our losses or discomfort, but didn't because we were busy with other things. In business, the "Bean Counters" take a jaundiced view of losses that could have been prevented with a little forethought and preparation.

The Federal Emergency Management Agency (FEMA) has prepared outlines for the development of emergency action plans. These are available from FEMA and the Internet (www.fema.gov). Local fire departments will also assist in developing such a plan. Another excellent source is *What to do When the Sky Starts Falling* by Janine Reid.[4] The process of developing an emergency action plan is similar to that of planning for a construction project. It generally starts with a vulnerability analysis. This phase in the emergency planning process is where the exposures present in a facility that could potentially be factors in an emergency situation are identified. Included in this analysis should be the issues identified during the safety analysis phase of the project. There, major services are identified, which if damaged, could result in interruptions of utilities, possible personal injuries, equipment damage, and environmental insults.

Tabulate the potential emergencies that could occur at or affect your facility and the people working in it. Develop a matrix of the emergency situations and the degree of probability, severity, and priority of response to each. Then conduct a "So What Analysis". This exercise in logic is where the possible outcomes of each event are identified and laid out as if ithey were the branches of a tree. Keep following the chain of thought by asking the question "So What?" and write down the alternatives and possible consequences of that decision, until there are no more reasonable alternatives which have not been addressed.

Scenario: Work is to be performed in a building with a CO_2 fire suppression system.

The potential hazard here that could degenerate into an emergency is the activation of the system resulting in an oxygen deficient atmosphere. What could go wrong? How should we prepare the contractor expected to work in the area?

---

[4] Janine Reid, *What To Do When The Sky Starts Falling: A Guide to Emergency Planning for the Construction Industry,* John Wiley & Sons, 1999.

Concern – CO2 Fire Suppression System Activation

Possible Scenarios

1. 30-second Pre-activation Warning Goes Off

Occupants rush to leave building	**So What?**
i. Trip over clutter in aisle ways	Ensure exits are not blocked
ii. Can't reach exit in 30 seconds	Emphasize housekeeping
	Train on building floor plan

2. No Warning Prior to Activation

CO2 is released	**So What?**
i. Occupants hear CO2 rush	Train on system warnings
ii. Occupants need air	Provide SCBA packs and training.
iii. CO2 whiteout occurs	Place strobes at exits
iv. Strobes fail without power	Emergency battery on strobes
v. Trip over clutter in aisles	Emphasis on housekeeping

Incidentally, this was a real case where work in a building with a CO2 system did go off following the de-energization of all the electrical systems in the building. The lights went out, the CO2 went off, a whiteout occurred, 8 of the 13 people in the building made it out. The air packs had been consolidated in another building. The clutter in the aisles blocked the egress of the other five individuals.

This "So What?" process is a remarkably effective means of evaluating any scenario because it develops an awareness of potential consequences without falling prey to the serious error of overlooking some basic issue which the participants might have assumed someone else was addressing or when following a predefined checklist. Individuals with the expertise to deal with the specific aspects of the identified problems should be invited to participate in developing the plan. Having these individuals involved in the development of response plans will also make them much more useful when they are called upon to respond in an emergency situation.

Analyze each potential emergency from beginning to end. Consider what could happen following an incident triggered from your site. How would your organization deal with the situation?

a. Who is responsible for notifying the media and who will talk to them?
b. Who will call the family of an injured employee?
c. Which local, state, or federal government agencies may need to be notified, and who will do so?
d. Who is responsible for notifying employees and who is the back up?
e. Your receptionists are your first line of defense; who is responsible for briefing them?
f. Do they know whom to contact within the company if they start to get many calls of a certain type, such as rumors?

g. Who will communicate your side of the story to the individuals important to the support of your operation (banker, bonding company, owner/developer, etc.)?

h. Has an incident/accident investigation team been established to begin the process of collecting statements and controlling the site as soon as immediate personnel issues have been addressed and before the production people start "Cleaning things up"?

While going through this process you should also identify what might be done to prevent the occurrences from materializing. Collect all the information that could potentially be of value when revising your safety program and the information that could be incorporated into your site safety orientation. The site safety orientation is an excellent opportunity to communicate with your contractor personnel the fact that you have an emergency plan in place and to identify any potential safety hazards on the job site. This is also the opportunity to advise individuals working on the job site what is expected of them were an event to occur. The information you convey may be as simple as advising the craftsmen that on the first Thursday of each month at 10 A.M. the sounding of emergency warning sirens is only a test and not to evacuate the site.

Although it is almost impossible to deal perfectly with all issues in a crisis, by anticipating as many of the events that could go wrong as you are able, you may keep a severed power line from damaging your company's reputation. There are few if any assets on your balance sheet worth your company's reputation in regard to getting future business.

IV.   Emergency Management Plan Documentation

At some point you must commit planned responses to emergency situations to paper. This will be your plan. However, before you start . . . think back, when was the last time you referred to the policy documents on your shelves? Will your first thought in the event of an emergency be to pull down the emergency documentation? Probably not. I do not advocate a "huge" plan, but it should be detailed enough for a new project engineer to understand the overall philosophy and the specific role of each predetermined player. In most cases, a succinct, well-defined document will serve the purpose very well. Your plan should include the following basic components:

A. Executive Summary

The executive summary provides a concise overview of the purpose of the plan, the facility's emergency management policy, authorities and responsibilities of key personnel, the types of emergencies which have been anticipated, and where response operations will be managed.

B. Emergency Action Plan Elements

This section of the plan outlines the program core elements and immediate response requirements.

    a.   Direction and control
    b.   Communications
    c.   Life safety
    d.   Emergency response procedures
    e.   Property protection
    f.   Community outreach
    g.   Recovery and restoration
    h.   Administration and logistics.

C. Support Documents

    a.   Emergency contact list
    b.   Employee list
    c.   Building and site maps
    d.   Safety information

Safety statistics such as days since the last lost time incident, periods worked without lost time injuries, and other information which highlights the organization's successful safety record should all be readily available. This is very important information when dealing with the media to establish that an incident that occurred is an anomaly and not a routine part of business.

    e.   Awards

Identify any awards or recognition for any of your work that you may have received. Identify everything positive that your company has done, both professionally and in the community. This is the information that will help put you into a positive light.

Clearly, more items may be included in your plan. However, this will give you a good start. Once the plan is in place, it is important to keep it current. Hold someone accountable to update the plan, phone numbers, accident statistics, awards etc. on a quarterly basis and make sure your key players know what is expected of them.

V.   Managing an Emergency Situation

Someone must be in charge during an emergency or it will surely develop into a crisis situation. Following an occurrence, but before the Crisis Management Team is assembled and has arrived on site, the media may have already arrived at your project site gate demanding to know why the ambulances have been called. The site superintendent may be the only person in authority present. He should know enough to only deliver a statement, which could be scripted as follows:

"My name is Bill Adams. I am the superintendent of Able Construction Company. As you are aware we have just experienced an accident that has apparently resulted in power interruption to the neighborhoods in this area and we have had two workers injured. Due to the rush associated with this emergency I do not have any detailed information at this time. Please give me 45 minutes to gather some facts for you. In the meantime, please stay in this safety area until I return."

Such a statement will show reporters that the company is not stonewalling and is willing to release information as soon as it is available. Remember "No Comment" implies guilt. Denying media access to information may convey the appearance that you are guilty or hiding something. This would certainly be reflected in media coverage of the situation. So, don't stonewall the media.

A. Roles and Responsibilities

Should an incident occur at your facility the individuals in your organization should be aware of their respective roles, what they should do, and who is responsible for managing the situation. In addition to the host of support roles, there are a few key players who must be appointed and prepared for their roles in an emergency situation.

1. Team Leader

The most important member of the emergency action team is the team leader. This person has the responsibility for controlling the situation. He must grasp the situation, and muster the resources necessary to deal with the occurrence. The team leader must be an individual able to resolve problems quickly and effectively. This person should have access to and the confidence of the CEO, President, and other top managers in the company. The team leader should be a participant in the corporate strategic planning effort and have a proven reputation for motivating and performing well with all functional managers. The team leader should be able to call on any member needed to implement the plan.

2. Spokesperson

The spokesperson is perhaps the most visible member of the team. The spokesperson represents the company in the public eye and will disseminate information about the situation to reporters, customers, and employees. The best candidate for this position is an individual who already is comfortable speaking to reporters or outside groups on behalf of your firm. If such a candidate is not available, now is the time to appoint and train a primary person and at least one alternate. Choosing the appropriate primary and alternate spokesperson(s) early is vital.

A common mistake made by unprepared companies faced with having to address the public is to automatically assume the CEO should be the spokesperson. The spokesperson should be that person who will best present, explain, and/or defend the company's position.

Once a spokesperson is selected, the responsibilities of this individual must be clearly delineated and communicated to the other members of the management team. A plan should be developed and explicitly document what is and what is not appropriate to say in an emergency. Issues such as timing, length and content of the statements, and even appearance of the speaker should be considered prior to the occurrence of an incident. A media consultant may even be considered to provide advice in grooming the spokesperson.

Don't forget your secretaries, telephone exchange operators, and receptionists. They are also your corporate spokespersons. Make sure they know what to say and how to say it, if they are expected to say anything at all.

VI.   Controlling the Message

The inability to communicate your message skillfully during a crisis can prove disastrous. The first few hours of an emergency are typically highly charged with rumors, speculation, and emotion. How you handle the first few hours will set the tone for the remainder of the time you are under public scrutiny. There are two sides to every story; no one says you have to tell both of them.

If there has been an occurrence at your facility, it is always better to initiate and release the news yourself. This allows you to choose how you word the situation, what you are doing to address the situation, and what action you plan to take. The alternative is that the news media will hear about it from a different source with a negative spin on it. Controlling the message by informing the public of the incident is an opportunity to demonstrate your leadership. Respond decisively and with conviction your abhorrence that such a crisis could have occurred and express sympathy for the individuals and families who may be have been affected.

Here is where you pull out your emergency Management Plan Information kit. In construction, where the potential crisis may be a serious personal injury, you should have all the latest and most pertinent safety records of your operation as of a reasonably recent date. This information will enable you to quickly issue statements that say things like:

- "Prior to today's mishap, our company has not lost one work day due to an accident in the last two years."

- "We are the leader in the field in terms of safety, as demonstrated by the following commendations from the Occupational Safety and Health Administration. . . ."

- "The equipment that malfunctioned and caused the trouble was last serviced and certified for use one month ago."

Such messages are important to get out early in a crisis. They demonstrate that you are concerned about safety and have taken a proactive role in the past to protect your employees. You are not hiding any truths, but presenting yourself in the best light and informing the concerned public that you are not a negligent employer. How long will it take to prepare this information if it is not ready before a crisis situation? How long will it take your company's communicator to gather such information and be prepared to issue such a message? Will that individual even have the time to do so when faced with dealing with a crisis?

VII.   Communicating with the Media

Dealing with the media is an area where facility owners and contractors often seem to get into trouble. Organizations that do not take the time to understand the media, learn its needs, and become comfortable with its members will probably get poor coverage or, worse yet, inaccurate, one-sided coverage. The penalty for poor communication is severe because there is no practical recourse to set the record straight. You can only make a good first impression once. Remember that the reporters are the individuals who control the message which finally reaches the public.

If you are prepared and have rehearsed, you stand a better chance of getting your side of the story to the public. Whether you find yourself in a one-on-one interview or at a press conference, facing either friendly or hostile media, honesty is of paramount importance. Being less than honest with the media or failing to respond to their needs will undoubtedly escalate your crisis. Failing to be communicative will destroy your present and future credibility with the media.

DO

Tell the truth – Clearly state the bottom line, no matter how bad it is. Be prepared to state what happened, when, where, how, who was affected, the extent of the damage, and cause, if known. Reporters will find it out anyway, so be honest and accurate when giving information. This doesn't mean you have to give every detail, but be truthful.

Be accurate – Don't destroy your credibility by speculating or failing to confirm your facts. Your information should come from a reliable source and you should understand the details thoroughly

Respond quickly – Be the first to release your information. If there is bad news, release it yourself before a reporter digs it up and tells the world. If you don't, the wrong story will be told and that is tough to erase. Don't force the press to ask a competitor or disgruntled employee to tell their version of your story.

Emphasize the positive and communicate your corporate message — Remember to emphasize the good safety measures taken, the minimal damage done because of the good team work of your employees, and what the company is doing to minimize the effect of the emergency on the public.

Stay away from liability issues — Do not talk about who is responsible; do not make any accusations; do not give out any company or individual names. Whatever you say will become part of a legal issue, so be as general as possible.

Make sure the reporters know who the spokesperson is — The corporate spokesperson should be the only one authorized to disseminate information outside of your company. That individual should be available to answer questions. It is very important that you speak with only one voice. Keep in mind that no information should be released without upper management approval.

DO NOT

Say anything "off the record" — If you don't want it used, don't say it. During a crisis, assume that everything you say to the media is very much on the record.

Say "no comment" — This is considered an admission of wrongdoing, ignorance, or arrogance. If you don't know the answer to a question, tell the reporters you don't know, but will try to find out. If you have nothing to say, say so and say why. "That's the first I have heard about it. I'd like to check into it before responding."

Get trapped into predicting the future — If the situation is complex and it will take days to determine the full extent of the damage, tell the reporters that the company will resume work on the project as soon as possible.

DO NOT wear sunglasses when being interviewed — You will be perceived as being shifty and hiding something.

Having worked on your communication skills prior to a crisis will be one more factor in your favor. Many companies are using public relations or media consulting firms to create simulated, but realistic conditions in which managers are subjected to real-life interviews to practice their skills. The interviews are taped, then played back and analyzed. Such drills enable people to become more aware of how to communicate with the public and the media.

VIII. Tips

Always know what your most important point is and lead with that. Keep focused on what you wish to communicate and don't be drawn off on tangents. Whenever you respond to a question make sure that you weave your main message into the answer. This is especially important for broadcast responses where the electronic media is looking for a good 30-second sound "bite."

One lesson in communication with the press is that people tend to remember the answer and not the question. So, give the answer you want to give and don't pay too much attention to the question. If you are uncomfortable with that concept, take the question and reshape it into one that serves the purpose of communicating your main point to the public.

Try to use graphic metaphorical imagery in making an important, statistical point. For example don't talk in terms of kilowatt hours; say " . . . enough electricity to light up a city the size of Springfield for a year." Make sure the audience can relate to the information.

Don't allow the interviewer to put words into your mouth. If he or she says, "What you mean to say is . . ." (if that is not what you meant to say), respond by saying, "No, that's not what I said, nor what I meant to say. What I said was . . ."

Be cautious of red-flag questions that begin with

- Isn't it true that . . . ?
- Aren't you really trying to say . . . ?
- How do you respond to . . .?
- Are you aware of . . .?

Before responding to a negatively worded question turn it into a positively worded one of your choosing. By restating the question you give yourself a bit more time and it ensures you are answering the question which was asked.

Some interviewers intentionally ask convoluted questions, or multiple questions in one breath, to try to confuse or fluster their interviewee. Be patient, let the interviewer finish; then begin your response by saying, "You've asked several questions, which I'll be glad to answer. Let me begin with your main point. . ." Then <u>define</u> what the main point is or what you want it to be.

You should never begin a response before a reporter has finished the question. This ensures that you have heard the whole question and it gives you the time to develop a response. If an interviewer continually interrupts you, don't be thrown. The more he or she talks the more time you have to work on your response. One reason why politicians often begin to answer a question with pat phrases is that it gives them an extra second or two to think before answering. Kennedy's "Let me say this about that," Nixon's "Let me be perfectly clear," Ferraro's "Let me suggest," and even Reagan's "Well, there you go again" are examples of how skilled politicians stall for time. Remember, the point of being in front of the news media is to communicate <u>your</u> point, don't be drawn off course by the media trying to make the situation more than it is. In crisis communication, what you say and how you say it are key to how you are perceived by the public.

Once you have made a statement, circulate a copy of the media statement to your employees to keep them informed. The media can and will talk to them. It is important that everyone understands the message being communicated to the media. At the conclusion of your statement, mention that you will have an update within a certain period of time. This lets the media know that you will continue to communicate with them.

IX.   Summary

By planning for circumstances which might go wrong, a company demonstrates that it is proactive and has taken steps to minimize the impact of an incident and to maintain control. An effective emergency and crisis action plan which is already in place when an incident takes place will minimize the impact of a crisis situation on your organization, enable normal operations to resume more quickly, and help protect the company's reputation. Additionally, being prepared for an emergency situation will reduce the need to deal with general administrative issues. Knowing in advance where the "flashlights are located and having core activities delegated" enables the emergency management team leader to focus on the root cause of the situation.

Begin to develop your plan today. Given the nature of construction and limited control over contracted services and given enough time, an accident will occur. Your organization must have guidelines defining who should do what when an emergency situation occurs. An organization that has such a plan will be in a position to respond quickly.

Contractors who have a crisis management plan demonstrate that they have done their long-range planning and are not satisfied with just dealing with problems that appear at their door each day. Owners should consider this in their technical evaluation of prospective contractors. This is a realistic acknowledgment that they see their business within construction as a potential high-risk industry and that undesirable incidents may occur even with the best safety program.

Emergencies don't discriminate between large or small contractors. However, they do seem to occur more frequently at those organizations that are not quite as prepared as others. Every firm involved in the construction industry should have an emergency management plan and be prepared to deal with a crisis when it occurs. Companies prepared to respond to an emergency situation are more likely to make everyone look prepared and in control, while a company without a plan may appear incompetent, inept and poorly managed.[5] Years of good public relations work can be undone in a single day. The time to prepare is now.

---

[5] An excellent resource in the area of emergency action planning and media training for the construction industry is Janine Reid, founder of Janine Reid Group, Inc., 1950 Jasmine Street, Denver, CO 80220, Phone (303) 322-3211, Fax (303) 316-3211. Author of "What To Do When The Sky Starts Falling" and numerous articles on the subject of crisis management.

Chapter 15

Contract Closeout Documentation and Transfer of Responsibility

We have all heard the saying that a job is not done until the paper work is complete. This is also true in regard to construction. Imagine the following scenario: two years after the project completion party is over and the keys to the facility were handed to the company President, you are served with papers to appear for a discovery deposition regarding an accident that occurred to a contractor's employee at your site during construction. Will you know about that occurrence? What information will you have to deal with that inquiry? There is no penalty for frivolous lawsuits in the U.S.; anyone can file one. Do you have access to the following information to prepare for your deposition or if the case goes to trial?

Construction contact ES&H requirements
Personnel injury reports
Project safety statistics
Site-wide and subcontractor safety meeting minutes

Inspection reports and progress photos
Job safety assessments
Material safety data sheets (MSDS)
Employee orientations records
Toolbox talks and sign-off sheets
Weekly man-hour reports
Engineer field notes and project notebook
Notices of unsafe acts
Work permits
Disciplinary notices

The Navy adage that you cannot launch a ship until you have generated enough paper to sink it seems to apply to construction these days as well.

I.   Introduction

The orderly closeout of contracts will minimize administrative difficulties down the road. The retention of appropriate documents will be very valuable when preparing a legal defense for a lawsuit you might be served with in the future. Having access to the right documentation will certainly minimize future headaches.

II.   Contract Closeout

Establish a process to alert you to the fact that a contract is nearing completion. (See Form 15.1.) In addition to being a reminder to verify that the compilation of needed documentation and information is completed, it will also give you the opportunity to ensure the contractor has returned any equipment or material that he might have borrowed before he leaves the site. As you may be well aware, once the contract is closed and any remaining retention is released, your leverage is gone. If you require anything of the contractor at that point you may be out of luck.

A. Contract Compliance

Let's approach this in a logical and orderly manner and deal with first things first. What deliverables have been specified in the contract language that relate to safety?

- Man-hour reports
- Personnel injury reports

As you have been maintaining project safety statistics and have a complete file of all accident reports by date, you certainly don't want a contractor to slip out and leave you with gaps in your records. The absence of man-hour information will lead to inaccuracies in your safety statistics.

B. System Acceptance Certifications

Have all building and system-related documentation been completed and delivered? Are the following tests and documents on file?

- Operating system tests completed
- Acceptance tests witnessed and documented
- Operations manuals and warranty information delivered
- System training sessions completed
- As-built drawings completed and submitted

What possible relationship do acceptance tests and operations manuals have with safety? Surely this is the responsibility of discipline engineers or the project manager. The fact is this information has a direct bearing on the safe operation of the completed facility. You need not personally keep these records, but you should have assurance that it is on file.

After completion of the facility, operation manuals can be a useful information source for operations personnel to develop their plant operations safety program. Operation manuals are also useful when developing the facility safety inspection checklist.

C. Loaned Equipment and Permits

What does the contractor have on loan that you want returned? Although it is not generally the custom to loan equipment to contractors, there are certainly situations where it is expedient to do so, such as the loan of two-way radios. Pagers are another means of locating specific personnel.

- Site access identification cards
- Radios and pagers
- Ladders
- Keys
- Open permits have been closed
- Lockout locks have been removed

Your greatest leverage over the contractor to get him to return this material is before making that final payment, release of retention or giving a letter of recommendation to his next customer. This is the time to make sure that everything you have loaned is returned.

D. Safety Program Documentation and Retention

In today's litigious society there is little need for a discussion regarding why we would wish to retain project-related safety records. The challenge is to determine what should be retained. Your corporation may have guidelines in this regard and your corporate legal council will surely have an opinion. The following are questions you will want to ask:

What information should be retained?
Where should it be stored?
How long should it be retained?
Who will know where to find it?

In my experience the collection and retention of the following information has been useful.

1. Safety Program Documentation

Information that will be requested of you first will be that related to your safety program and its structure, i.e., program documentation, permit and inspection forms, training syllabuses, safety posters, etc. After the completion of every successful project, corporate affiliates or other organizations that have heard of my program's successes asking for information and copies of the project documentation have approached me. Their intent was to emulate my success on their own construction projects. This information has some shelf life, since at the conclusion of a job one looks back and sees a number of things that could be improved on the next job.

The historical information that relates to the implementation of your program should include, at a minimum, the following information, all of which are evidence of the effective execution of your safety program.

- Joint inspection reports
- Weekly site safety audits and progress photos
- Safety meeting minutes
- Job Safety Assessments (JSA)
- Toolbox talks and sign-off sheets
- Employee orientation outline and attendance records
- Approved safety variances
- Enforcement/reinforcement documentation
- Safety notices

2. Safety Statistics

At the conclusion of a project there are always closeout meetings and presentations to advise the clients how their money was spent and the virtues of the product they are receiving. Without exception, safety statistics are an integral part of that information (Chapter 13 — Safety Statistics and Reports).

Man-hour reports and summaries
OSHA No. 200 – Log of Summary and Occupational Illness and Injuries
OSHA No. 101 – Supplementary Record of Occupational Injuries and Illnesses
Commendations and Citations
Copies of articles that reported on your project's successes

3. Accident Reports and Personal Injury Information
Do you have a complete file of all accidents reported and investigated? These can be filed either by contract or chronologically. After all the effort we have put into our program to date, this should be a relatively small file.

- Employer's first report of injury
- Employee's first report of injury (if state requires)
- Documentation that the employee's report of injury was given to the company within 24 hours of an injury (if state requires)
- Supervisor's accident investigation report
- All other claim-supporting documentation

4. Other Safety-Related Information
Did your project engineers keep daily logs of their field activities? Where are they? Are they being taken home for their kids to draw on? These sorts of documents have been quite useful in mounting an effective defense to demonstrate the active involvement of field engineers in overseeing safety on the construction site. The following would round out your files:

Information on Site Conditions and Contractor Performance

- Field engineer daily logs
- Job progress photos/video tapes
- Field inspection reports and related hazard abatement information

Work and Safety Planning Information

- Safety meetings — Site wide and contractor specific
- Safety related planning documentation and correspondence
- Job-safety assessments and sign-off sheets
- Audit results and documentation
- Resolution of audit results

Monitoring Records

- Exposure and medical records (should be maintained 30 years)
- Environmental test results (30 years)
- Noise exposure records (3 years)

This may be quite a bit of material depending on the size of your project. However, if you collect this information as the project evolves, your files will then constitute the records you need to retain when the project is complete. If you haven't started your project yet, you now have the outline for a good records index.

E. Where Should It Be Stored?

The safest place for this information is in long-term retention. It is not underfoot and is therefore less likely to dumped in a housekeeping effort. Whatever location you select, you should have confidence that it will not be thrown out inadvertently. Perhaps marking the boxes with prominent labels that direct the handlers to call you or the legal department before they are eventually discarded might be useful.

F. How Long Should It Be Retained?

There are numerous requirements for the retention of employee-related safety records. OSHA records are required to be kept for five (5) years per federal guidelines; however, individual state statutes and regulations do vary. Other routine files must be retained under the Department of Labor, 29 CFR Part 516 and state requirements. These retention periods are generally short – just a few years. Workers' compensation requirements for retention periods vary by state, but are usually between one to three years.

The aforementioned requirements guide the minimum retention of employee-related documents, but from the perspective of developing a defense in the case of litigation for construction type injuries, the statute of limitation is generally six (6) years. Occupational illnesses may take more time to evolve. Therefore, records of work environment conditions monitoring air pollutants, noise, radiation and the like should be retained for thirty (30) years.

G. Who Will Know Where To Find It?

This is perhaps the most challenging issue. I suggest copies of the document inventory listing be forwarded to the Legal Department and to Procurement to be filed with their contract-related records, then to the attention of whomever a potential claim would be presented. They will know where any safety-related information can be found.

We are almost done.

III.   Final Steps

As the project nears completion and the last of the contractors demobilize, you need to advise all those support organizations upon whom you depended for emergency response, technical support, and other services that the project is coming to an end and they need not be on stand-by any longer.

You also need to meet with the owner to advise him of the steps you have taken to close out the project. The owner is going to be holding the bag in the event there are claims in the future. Does he know where you have archived the files and what is in the files?

As construction draws to a close and the owner takes over the new facility, is it clear who is responsible for safety?

IV.   Conclusion

If you have access to the information outlined in this chapter, you have been managing an effective safety program and you should be able to sleep comfortably at night.

Now, on to your next successful and safe project.

Form 15.1
Inspection & Acceptance Memorandum

Subject: _____

    Contract No._____

    Contractor: _____

Contractor Statement of Completion

An inspection of the work performed under the subject contract indicates that its as-built documents and operation/maintenance manuals have been submitted to the project manager. The execution of the work ☐ **did,** ☐ **did not** involve Contractor use of Laboratory owned materials or equipment and a complete property settlement exists at this time.

Contractor _____ Date _____

Conformance with ES&H Requirements

The subject facility is considered to be satisfactory and is in general conformance with the applicable ES&H requirements.

ES&H _____ Date _____

Conformance with Utilization Requirements

The subject facility is considered to be satisfactory and in general conformance with the utilization requirements that were furnished.

Requesting Division _____ Date _____

Conformance with Design Drawings and Specifications

The subject facility is considered to be satisfactory and in conformance with the contract drawings, specifications and approved changes. All contractually required as-built documents and operation/maintenance manuals have been received.

Project Manager _____ Date _____

Acceptance for Operation and Maintenance

The subject facility is considered to be satisfactory and is accepted by Plant Facilities and Services for operation and maintenance.

Facilities and Svcs. _____ Date _____

The work is hereby accepted as of _____. This acceptance does not in any way waive the provisions of the contract regarding guarantees against defects in materials and workmanship.

Procurement _____ Date _____

EPILOGUE

Managing construction site safety is an integrating activity and is done best when the project manager has a balanced outlook on all elements essential to the success of any project: manpower, equipment, material, cost and schedule planning, as well as safety. Projects are typically subject to numerous conflicting demands and pressures on all involved personnel. However, to implement a successful safety program requires commitment and unwavering support of the basic tenet of wanting to ensure that the workforce is fit and able to report to the job site each day free of injuries.

For ease of explanation and understanding, many of the chapters in this book deal with individual aspects of construction site safety. This is not to imply that these aspects stand alone, or can be pursued independently. For example, controls cannot be established for hazards unless the hazards have been identified, and controls are of little value unless they are implemented. One cannot control or eliminate hazards without planning, training, monitoring, and coordinating, nor can one report without monitoring. Thus, I urge you to integrate safety into all aspects of planning and execution of your projects.

Many projects get into trouble and some become utter failures; that is, time, money, and effort are spent without achieving the objective of establishing an injury-free project. While the reasons claimed for trouble or failure are varied, they generally boil down to inadequate safety management.

Sometimes project managers don't understand that managing safety is one of their responsibilities and is just as important to their project's success as is work planning, scheduling, and coordination of the work itself.

Sometimes project managers fail to exercise their prerogatives. They may, for example, assume the construction manager or general contractors will enforce safety of their own accord. They may not ask for timely, meaningful safety reports or consider monitoring the contractors' implementation of safety and are not aware of the degree to which unsafe work practices are present until some major tragedy has occurred.

Sometimes project managers dupe themselves into thinking that because the contract language has a clause that requires their subcontractors to follow all federal, state, and local safety regulations there is no reason for them to be concerned about safety.

Project managers and field engineers are in a position of trust. They are not only the focal points for their project; they are also the stewards of all activities that take place on their projects. As a project manager or field engineer, in the eyes of your employer, you are responsible for the work performed by subcontractors, as if you had done the job yourself.

It is a pleasure to be part of projects with successful safety programs. An ineffective safety program is a tragedy. I hope that this book will help you, your project team, and your employers better understand how to implement your safety program and give you the satisfaction derived from a successful construction safety program.

Index